Green Energy and Technology

Climate change, environmental impact and the limited natural resources urge scientific research and novel technical solutions. The monograph series Green Energy and Technology serves as a publishing platform for scientific and technological approaches to "green"—i.e. environmentally friendly and sustainable—technologies. While a focus lies on energy and power supply, it also covers "green" solutions in industrial engineering and engineering design. Green Energy and Technology addresses researchers, advanced students, technical consultants as well as decision makers in industries and politics. Hence, the level of presentation spans from instructional to highly technical.

Indexed in Scopus.

More information about this series at http://www.springer.com/series/8059

Pratima Bajpai

Developments in Bioethanol

Springer

Pratima Bajpai
Pulp and Paper Consultants
Kanpur, Uttar Pradesh, India

ISSN 1865-3529 ISSN 1865-3537 (electronic)
Green Energy and Technology
ISBN 978-981-15-8778-8 ISBN 978-981-15-8779-5 (eBook)
https://doi.org/10.1007/978-981-15-8779-5

This Springer imprint is published by the registered company Springer Nature Singapore Pte Ltd.
The registered company address is: 152 Beach Road, #21-01/04 Gateway East, Singapore 189721, Singapore

Preface

Disadvantages of fossil-fuel-derived transportation fuels (greenhouse gas emissions, pollution, resource depletion, and unbalanced supply–demand relations) are strongly reduced or even absent with bio transportation fuels. Of all biofuels, ethanol is already produced on a fair scale. It is biodegradable, produces slightly less greenhouse emissions than fossil fuel (carbon dioxide is recycled from the atmosphere to produce biomass), can replace harmful fuel additives (e.g., methyl tertiary butyl ether), and produces jobs for farmers and refinery workers. It is easily applicable in present-day internal combustion engine vehicles (ICEVs), as mixing with gasoline is possible. Ethanol is already commonly used in a 10% ethanol/90% gasoline blend. Adapted ICEVs can use a blend of 85% ethanol/15% gasoline (E85) or even 95% ethanol (E95). Ethanol addition increases octane and reduces CO, VOC, and particulate emissions of gasoline. And, via on board reforming to hydrogen, ethanol is also suitable for use in future fuel cell vehicles (FCVs). Those vehicles are supposed to have about double the current ICEV fuel efficiency.

Ethanol production and use have spread to every corner of the globe. As concerns over petroleum supplies and global warming continue to grow, more nations are looking at ethanol and renewable fuels as a way to counter oil dependency and environmental impacts.

Growth in ethanol production is driven by United States, Brazil, followed by China, expanding markets in India and Thailand. IEA predicts that China's share of global production will increase in the next few years, but the United States and Brazil will still account for 80% of worldwide output. While the vast majority of ethanol is consumed in the country in which it is produced, some nations are finding it more profitable to export ethanol to countries like the United States and Japan. High spot market prices for ethanol and the rapid elimination of MTBE by gasoline refiners led to record imports into the United States in the last few years. More than 500 million gallons of ethanol entered through American ports, paid the necessary duties, and competed effectively in the marketplace. The increased trade of ethanol around the world is helping to open up new markets for all sources of ethanol.

The sustainable production of bioethanol requires well-planned and reasoned development programs to assure that the many environmental, social, and economic concerns related to its use are addressed adequately. The key for making ethanol competitive as an alternative fuel is the ability to produce it from low-cost biomass. Many countries around the world are working extensively to develop new technologies for ethanol production from biomass, from which the lignocellulosic materials conversion seem to be the most promising one. This e-book provides an updated and detailed overview on Developments in Bioethanol. It looks at the historical perspectives; chemistry; sources and production of ethanol; and discusses biotechnology breakthroughs and promising developments, its uses, advantages, problems, environmental effects, and characteristics. In addition, it presents information about ethanol in different parts of the world and also highlights the challenges and future of ethanol. It highlights the evolution in bioethanol development from first-generation production to the futuristic fourth-generation bioethanol production, the various constraints and challenges involved, and the scope for further development. This book is an extension of the earlier Springer Briefs—Advances in Bioethanol published in 2013.

Kanpur, India

Pratima Bajpai

Contents

List of Figures

List of Tables

Chapter 1
General Background and Introduction

Abstract Biofuels are a potential renewable energy source to replace fossil fuels, particularly because of much reduced greenhouse gases emissions. The use of ethanol as an alternative motor fuel has been progressively increasing around the world. Domestic production and use of ethanol for fuel can decrease dependence on foreign oil, reduce trade deficits, create jobs in rural areas, reduce air pollution, and reduce global climate change carbon dioxide build up. Ethanol, unlike gasoline, is an oxygenated fuel that contains 35% oxygen, which reduces particulate and NOx emissions from combustion. Unlike fossil fuels, ethanol is a renewable energy source produced through fermentation of sugars. This chapter presents general background and introduction to developments in bioethanol.

Keywords Biofuels · Bioethanol · Fossil fuel · Greenhouse gases · Oxygenated fuel

Energy consumption has increased steadily over the last century as the world population has grown and more countries have become industrialized. Crude oil has been the major resource to meet the increased energy demand. Campbell and Laherrere (1998) used several techniques to estimate the current known crude oil reserves and the reserves as yet undiscovered. They predicted that annual global oil production would decline to approximately 5 billion barrels in 2050. Because the economy in the United States and many other nations depends on oil, the consequences of inadequate oil availability could be severe. So, there is a great interest in exploring alternative energy sources (Wyman and Hinman 1990; Lynd and Wang 2004; Herrera 2004; Tanaka 2006; Dien et al. 2006; Sun and Cheng 2004; Yacobucci and Womach 2003; Chandel et al. 2007; Gray et al. 2006; Kheshgi et al. 2000; DOE 2007a; Badger 2002; EBIO 2006; Bajpai 2007; Ghose and Ghose 2003; Hazell and Pachauri 2006; Hazell and von Braun 2006; Joseph 2005; Launder 1999; Luhnow and Samor 2006; Paris Morris 1993; Nastari 2005; Nastari et al. 2005; REN21 2006; RFA 2006, 2017, 2019; Rosillo-Calle and Walter 2006; Demirbas 2007, 2009; Ward and Singh 2002; IPCC 2001).

P. Bajpai, *Developments in Bioethanol*, Green Energy and Technology,
https://doi.org/10.1007/978-981-15-8779-5_1

The world is mainly dependent upon non renewable energy sources for generation of heat and power and transport. The major energy source is fossil fuel. These fuels, provide about 78.4% of the final energy consumption worldwide (Renewables 2017). Because of the increasing energy requirements and impacts of fossil fuels use on health and environment, there is an urgent requirement to explore other options (Hussain et al. 2017). Furthermore, depending on the current usage, the discovery rate of fossil fuels, shortly, will not match the utilization rate (Du et al. 2016). Biofuels emit lesser greenhouse gases. Therefore are a promising source for replacing fossil fuels. In addition, biofuels are produced from common biomass sources which are geographically more consistently distributed as compared to fossil fuels, allowing for an autonomous and secured supply of energy (Liew et al. 2014; Nigam and Singh 2011). Due to these reasons, there is an increased interest in biofuels by researchers. Several articles are being published in this area (Azadi et al. 2017).

The use of ethanol as an alternative motor fuel has been steadily increasing around the world (BP 2006; Jessel 2006; Moreira and Walter 2005). Domestic production and use of ethanol for fuel can decrease dependence on foreign oil, reduce trade deficits, create jobs in rural areas, reduce air pollution, and reduce global climate change carbon dioxide build up. Ethanol, unlike gasoline, is an oxygenated fuel that contains 35% oxygen, which reduces particulate and NOx emissions from combustion. Unlike fossil fuels, ethanol is a renewable energy source produced through fermentation of sugars.

Renewable Fuel Association (RFA) (2017) reported that the blending of ethanol in gasoline, reduced carbon dioxide-equivalent green house gas emissions from transportation by 43.5 million metric tonnes in 2016. This is equivalent of removing 9.3 million cars from the road for the whole year (RFA 2017). Use of bioethanol further reduces dependence on crude petroleum, which is generally imported from abroad, increasing energy security and expanding energy supplies. It is also helping to increase employment and triggering the economy in the countryside (Senthilkimar and Gunasekaran 2005; Sims et al. 2010). Brazil and the United States were the first bioethanol producing countries.

In the United States, the main raw material used is corn starch. In 1998, 6.4 billion litre of bioethanol was produced (Berg 1999). By 2007, output increased almost four times to 24.71 billion litre, by 2010 it had more than doubled to 50.41 billion litre. In 2013, production reduced slightly to 50.37 billion litre. However, in 2016, it increased again to 57.8 billion litre. At present, there are more than 200 biorefineries in the United States, having the capacity of producing about 60.64 billion litre of ethanol per year jointly (RFA 2017).

In Brazil, sugarcane juice is the major raw material used. The production reached about 13.5 billion litre in 1998 (Berg 1999).

In Europe, several countries are involved in bioethanol production. These are Sweden, France, Germany, Italy and the UK. European Union countries are jointly producing more than 2 billion L of bioethanol annually (Robak and Balcereket 2018).

Bioethanol is playing an important role in renewable energy particularly in the transportation sector. Governments in many countries are encouraging the production and utilization of bioethanol for reducing fossil dependency, reducing global warming impact, and activating the grass-root economy by stabilizing the income of farmers and creating employment in the local community.

Bioethanol is an ethanol produced from biomass. So, it is renewable. It has some benefits over petrol as fuel. As the biomass grows, it is able to consume as much carbon dioxide as it is formed during the combustion of bioethanol. This makes the net contribution to the green house effect zero. If use of bioethanol is encouraged, the rural economy would also receive a boost from growing the essential crops. Furthermore use of bioethanol in older engines may help in reducing the amount of carbon monoxide produced by the vehicle thereby improving air quality. Another benefit of bioethanol is that it can be easily integrated into the existing road transport fuel system. Bioethanol can be blended with conventional fuel up to 5%, without any requirement of modifying the engine.

The term biofuel is attributed to any alternative fuel that derives from organic material, such as energy crops (corn, wheat, sugar cane, sugar beet, cassava, among others), crop residues (e.g. rice straw, rice husk, corn stover, corn cobs) or waste biomass (for instance, food waste, livestock waste, paper waste, construction-derived wood residues and others). Of all biofuels, ethanol has been trusted as an alternate fuel for the future and is already produced on a fair scale worldwide. Most of the production is located in Brazil and the United States (ANP 2007; Hamelinck et al. 2005). Bioethanol is expected to be one of the dominating biofuels in the transport sector within the next 20 years (Hägerdal et al. 2006; Berg 2004; Paszner 2006).

In the year 2016, bioethanol, corresponded to ~73% of the 135.3 billion liters of biofuel produced. The largest producer is the United States, 59%, followed by Brazil, producing about 27% of the bioethanol worldwide (Renewables 2017; Branco et al. 2019). Bioethanol can be mixed with gasoline or can be used as a replacement of gasoline. The use of bioethanol in spark ignition engines shows several benefits when comparison is made with gasoline. Ethanol has a higher oxygen content, which encourages better combustion and reduced emission of exhaust gases, and a higher octane number. This would allow engine to operate at a higher compression ratio. Furthermore, use of vegetable biomass as substrate for bioethanol production would allow for recycling the carbon dioxide emitted during combustion. This would reduce the carbon dioxide emissions (Balat 2011; Sebayang et al. 2016).

Bioethanol can be also used as a platform chemical for producing numerous molecules (for instance diethyl ether, ethylene, propylene, acetaldehyde, and ethyl acetate), in beverages, pharmaceuticals, and cosmetics (Sarris and Papanikolaou 2016; Maity 2015). At the present time, most of the commercial bioethanol is almost of first generation as food crops are used as substrate: sugarcane in Brazil, corn in the United States, and wheat and sugar beet in the European Union. The major drawback of first generation bioethanol is the competition with the use of arable land for growing of food crops between biofuel feedstocks, thereby resulting in the increase of food prices (Branco et al. 2019; Manochio et al. 2017; Dutta et al. 2014; Bastos 2018).

Bioethanol can be utilised as a liquid fuel in internal combustion engines, either neat or in blends with petroleum (Table 1.1). Table 1.2 compares the energy content of bioethanol with conventional fossil fuels used for road and aviation transportation. In Brazil, ethanol blends are mandatory (E20 to E25) and anhydrous ethanol (E100) is

Table 1.1 Bioethanol-gasoline blends used in different countries

Blend	Country	
Max. 5% anhydrous ethanol, min. 95% petrol	Western Europe	Blends for regular cars
Max. 10% anhydrous ethanol, min. 90% petrol	USA, Europe	Blends for regular cars
Max. 15% anhydrous ethanol, min. 85% petrol	USA, cars >2000	Blends for regular cars
Max. 25% anhydrous ethanol, min. 75% petrol	Brazil	Blends for regular cars
Max. 85% anhydrous ethanol, min. 15% petrol	USA, Europe	Flex-fuel vehicles
Hydrous ethanol (~5.3 wt% water)	Brazil	Flex-fuel vehicles

Based on Walker (2010), Baeyens et al. (2015), Kang et al. (2014)

Table 1.2 Energy content of Fossil fuels and bioethanol

Fuel	Energy content, MJ/L
Gasoline (regular)	34.8
Gasoline (aviation)	33.5
Diesel	38.6
Autogas (LPG)	26.8
E100	23.5
E85	25.2
E10	33.7

Based on Walker (2010)

also available from thousands of filling stations. In addition, there are 6 million flex-fuel vehicles in Brazil and 3 million able to run on E100. Bioethanol now accounts for ~50% of the Brazilian transport fuel market, where gasoline may now be regarded as the "alternative" fuel. In Brazil >20% of cars (and some light aircraft) are able to use E100 (100% ethanol) as fuel, which includes ethanol-only engines and flex-fuel vehicles which are able to run with either neat ethanol, neat gasoline, or any mixture of both (Lynch 2006).

Today, the main reason for the interest in renewable biofuels is the possibility of obtaining a substantial reduction of noxious exhaust emissions from combustion, especially as statutory limits are becoming more stringent and more exhaust components are regulated. Wider use of a chemically simple fuel such as bioethanol will mean that there are fewer harmful effects on life and ecosystems. Using ethanol in place of gasoline helps to reduce carbon dioxide (CO_2) emissions by up 30–50% given today's technology (Table 1.3). In particular, people living in urban areas may in future appreciate the use of improved low-emission vehicles that have an agreeable smell, are smokeless and are propelled either by reformulated bioethanol, by bioethanol blended with gasoline, or by neat biofuels. How the air quality can be improved is something that is increasingly worth investigating, for the sake of people and the environment! Large-scale, sustainable, worldwide production and use of bioethanol from biomass resources will produce tangible significant benefits

Table 1.3 Ethanol facts (Environment)

Using ethanol in place of gasoline helps to reduce carbon dioxide (CO_2) emissions by up 30–50% given today's technology
In 2012, the 13.2 billion gallons of ethanol produced reduced greenhouse gas emissions from on-road vehicles by 33.4 million tons. That's equivalent to removing 5.2 million cars and pickups (comparable to the number of registered vehicles in the state of Michigan) from the road for one year
New technologies are increasing ethanol yields, improving efficiencies and allowing ethanol biorefineries to make better use of natural resources
Ethanol has a positive energy balance
The American Lung Association supports the use of E85, a blend of 85% ethanol and 15% gasoline for flex-fuel vehicles specifically designed to operate on this fuel
Ethanol reduces tailpipe carbon monoxide emissions by as much as 30%, toxics content by 13% (mass) and 21% (potency), and tailpipe fine particulate matter (PM) emissions by 50%
Ethanol is the oxygenate of choice in the federal winter oxygenated fuels program in cities that exceed public health standards for carbon monoxide pollution
Ethanol is rapidly biodegraded in surface water, groundwater and soil, and is the safest component in gasoline today.
Ethanol reduces smog pollution
Water usage in ethanol production is declining

for our growing and fast-evolving society, as well as for the earth's climate. Important environmental benefits could be achieved in the socio-economic development of large rural populations and the diversification of energy supply, in particular for the strategically vital sector of transport (Turkenburg 2000).

A life cycle analysis of ethanol production—from field to the car—by the United States Department of Agriculture found that ethanol has a large and positive energy balance. Ethanol yields 134% of the energy used to grow and harvest the corn and process into ethanol. By comparison gasoline yields only 80% of the energy used to produce it. Bioethanol does not add to global carbon dioxide levels because it only 'recycles' carbon dioxide already present in the atmosphere. Carbon dioxide gets removed from the atmosphere through photosynthesis when crops intended for conversion to bioethanol are grown (Fig. 1.1). Carbon dioxide is then released to the atmosphere during combustion. On the contrary, burning a fossil fuel such as petrol adds to global carbon dioxide as it releases new amounts of carbon dioxide that were earlier trapped underground for millions of years. Ultimately, dissimilar to oil, bioethanol is a renewable fuel, which intrinsically helps the environment by allowing to preserve other energy resources (EIA 1999, 2006).

The Advantages and Disadvantages of Ethanol Fuel are detailed in Table 1.4. Ethanol is a flexible transportation fuel that can be used in anhydrous form at 99.6 Gay Lussac (GL) as a blending agent in ethanol-gasoline blends, either directly or indirectly or as a primary fuel in neat hydrous form (95.5 GL). Ethanol makes an excellent motor fuel: it has a research octane number (RON) of 109 and a motor octane number (MON) of 90, which are higher compared to gasoline, which has a RON of 91–98 and a MON of 83–90. Ethanol also has a lower vapor pressure than gasoline (its Reid vapor pressure is 16 kPa versus 71 for typical gasoline), resulting in less evaporative emissions. Ethanol's flammability in air (1.3 to 7.6% v/v) is also

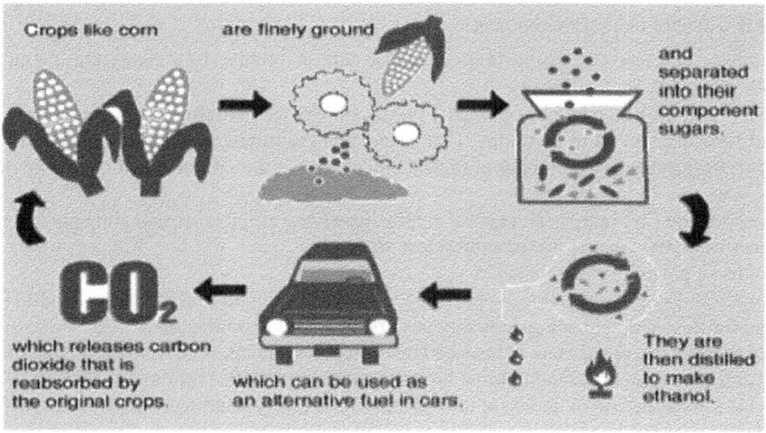

Fig. 1.1 Fuel ethanol process. Based on Launder (1999)

Table 1.4 Advantages and disadvantages of Ethanol Fuel

Advantages
Renewable, relatively safe fuel that can be used with few engine modifications
Can improve agricultural economies by providing farmers with a stable market for certain crops, such as maize and sugar beets
Increases national energy security because some use of foreign petroleum is averted
Nontoxic and biodegradable, quickly breaks down into harmless substances if spilled. Ethanol use reduces carbon monoxide and many toxic pollutants from the tailpipe of vehicles, making air cleaner
As it is made from crops that absorb carbon dioxide and give off oxygen it has a potential to reduce greenhouse gas emission and help maintain the balance of carbon dioxide in the atmosphere
Ethanol, unlike petroleum, is claimed to be a form of renewable energy that can be produced from agricultural crops such as sugar cane, potato, and corn It has many other advantages over gasoline at least it can be a permanent resource
Disadvantages
Ethanol has a lower heat of combustion (per mole, per unit of volume, and per unit of mass) that petroleum
Large amounts of arable land are required to produce the crops required to obtain ethanol, leading to problems such as soil erosion, deforestation, fertiliser run-off and salinity
Major environmental problems would arise out of the disposal of waste fermentation liquors
Ethanol powered vehicles will have trouble starting up at low temperatures
Typical current engines would require modification to use high concentrations of ethanol
Harder to transport

Based on http://www.wisegeek.org/what-are-the-advantages-and-disadvantages-of-ethanol-fuel.htm
http://www.solarpowernotes.com/renewable-energy/ethanol-fuel.html

Table 1.5 Major uses of Ethanol

Beverage alcohal
Ethanol is often called 'drinking alcohol' as it is the prime ingredient in alcoholic beverages. Ethanol is the intoxicating substance in alcohol
Fuel alcohal
Ethanol can be used as a fuel for motor vehicles. Ethanol make a good fuel for cars because it reduces the emission of harmful gases such as carbon monoxide. Brazil are one of the leaders in the production of cars that run on ethanol. Over 20% of cars in Brazil are able to run on 100% ethanol fuel
Lightweight rocket-powered racing aircraft often use ethanol as rocket fuel
Ethanol is used in antiseptic and some antibacterial soaps and wipes. Ethanol is effective against viruses, fungi and most bacteria but is ineffective against bacterial spores
Industrial alcohal
As ethanol is soluble in water, it can be used in a variety of different products. These include paint, permanent markers, perfumes and deodorants. Ethanol may also be used as a solvent in cooking, such as vodka sauce
Ethanol is considered a 'feedstock' into the chemical industry as it is used to make other important chemicals

Based on Berg (2004)

much lower compared to gasoline (3.5 to 19% v/v), reducing the number and severity of vehicle fires (Goldemberg et al. 1993). On the other hand, when used as a neat fuel, ethanol has a lower energy density than gasoline (ethanol has lower and higher heating values of 21.2 and 23.4 MJ/L, respectively; for gasoline the values are 30.1 and 34.9 MJ/L) and cold-start problems exist as well (McMillan 1997).

Ethanol is also suitable for use in future fuel cell vehicles (FCVs) (DOE 2007b). Those vehicles are supposed to have about double the current ICEV fuel efficiency (Lynd 1996). Beginning with the model year 1999, an increasing number of vehicles in the world are manufactured with engines which can run on any gasoline from 0% ethanol up to 85% ethanol without modification. Many light trucks are designed to be dual fuel or flexible fuel vehicles, since they can automatically detect the type of fuel and change the engine's behaviour, principally air-to-fuel ratio and ignition timing to compensate for the different octane levels of the fuel in the engine cylinders.

Ethanol has three major uses: as a renewable fuel, as a beverage, and for industrial purposes (Table 1.5). Of the three grades of ethanol, fuel grade ethanol is driving record ethanol production in many countries. About 95% of all ethanol is derived by fermentation from sugar or starch crops; the rest is produced synthetically. The synthesis route evolves dehydration of hydrocarbons (e.g., ethylene) or by reaction with sulphuric acid, to produce ethyl sulphate, followed by hydrolysis. The production routes from biomass are based on fermentation or hydrolysis. According to F.O. Licht, synthetic alcohol production is concentrated in the hands of a few mostly multinational companies such as Sasol, with operations in South Africa and Germany, SADAF of Saudi Arabia, a 50:50 joint venture between Shell of the United Kingdom (UK) and Netherlands, the Saudi Arabian Basic Industries Corporation, BP of the UK as well as Equistar in the United States

Table 1.6 World ethanol production by country, in percent

USA	57
Brazil	27
Europe	6
China	3
India	2
Canada	2
Row	3

https://www.e-education.psu.edu/egee439/node/646

Fermentation ethanol is mainly produced for fuel, though a small share is used by the beverage industry and the industrial industry. The bulk of the production and consumption is located in Brazil and the United States. Fermentation technologies for sugar and starch crops are very well developed, but have certain limits—these crops have a high value for food application, and their sugar yield per hectare is very low compared with the most prevalent forms of sugar in nature: cellulose and hemicellulose. Suitable processes for lignocellulosic biomass therefore have room for much further development: a bigger crop variety can be employed, a larger portion of these crops can be converted, and hence larger scales and lower costs are possible. There is copious amount of lignocellulosic biomass worldwide that can be exploited for fuel ethanol production. According to United States Dept. of Energy, cellulosic ethanol reduces greenhouse gas emissions by 85% over reformulated gasoline. By contrast, sugar-fermented ethanol reduces greenhouse gas emissions by 18% to 19% over gasoline. Beyond added environmental benefits, cellulose-based ethanol could offer additional revenue streams to farmers for the collection and sale of currently unused corn stover or straw, for example.

In the United States, the common method of producing fuel ethanol is from starches, such as corn, wheat, and potatoes. The starch is first hydrolyzed to glucose before proceeding with the remaining process. In Brazil, sucrose, or sugar in sugarcane are most commonly used substrates. In Europe, the most common feedstock is sugar beets. Cellulose is being utilized in developing methods, which includes wood, grasses, and crop residues. It is considered developing, as conversion of cellulose into glucose is more challenging as compared to sugars and starches.

"The International Energy Agency (IEA) predicts that ethanol will constitute two-thirds of the global growth in conventional biofuels with biodiesel and hydrotreated vegetable oil accounting the remaining part (2018–2023). Global ethanol production is estimated to increase by 14% from about 120 bln L in 2017 to approximately 131 bln L by 2027. Brazil will accommodate fifty per cent of this increase and will be used to fill in the domestic demand" (FAO 2018).

World production of ethanol based by country is presented in Table 1.6 and Fig. 1.2. The United States produces the most ethanol worldwide, mainly from corn. Brazil is the next largest producer, mostly from sugarcane. Other countries, including

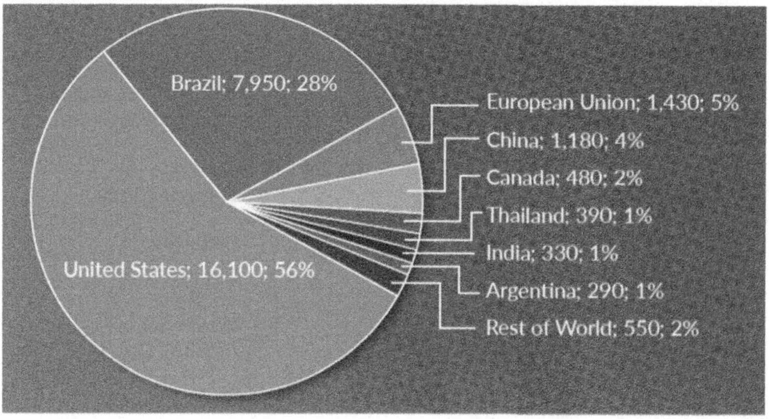

Fig. 1.2 2018 global fuel ethanol production by country. RFA (2019), reproduced with permission (Country, million gallons, share of global production)

Table 1.7 Predictions of the world bioethanol production (**a**) and consumption (**b**) by 2024

	Production	Consumption
USA	42	41
Brazil	31	29
European Union	7	8
China	7	7
India	2	2
Thialand	2	2
Others	9	11

Based on OECD-FAO (2015), Bušic et al. 2018

Australia, Columbia, India, Peru, Cuba, Ethiopia, Vietnam, and Zimbabwe, have also started to produce ethanol from sugarcane.

Table 1.7 presents predictions of the world bioethanol production (a) and consumption (b) by 2024.

Figure 1.3 shows ethanol production in United States by feedstock type.

"Corn is the major substrate for *United States* ethanol, and sugarcane is the main source of ethanol in Brazil. In the United States, the record production of biofuels is attributed in part to high oil prices, which pushed many large fuel companies, including Sunoco, Valero, Flint Hills, and Murphy Oil, to produce ethanol. High oil prices were also a factor in Brazil, where every third car-owner drives a flex-fuel vehicle that can run on either fossil or biofuels. Although the United States and Brazil are the world leaders in ethanol whereas the largest producer of biodiesel is the European Union. But, some European countries may switch from biodiesel to ethanol as a recent report from the European Commission shows that ethanol crops

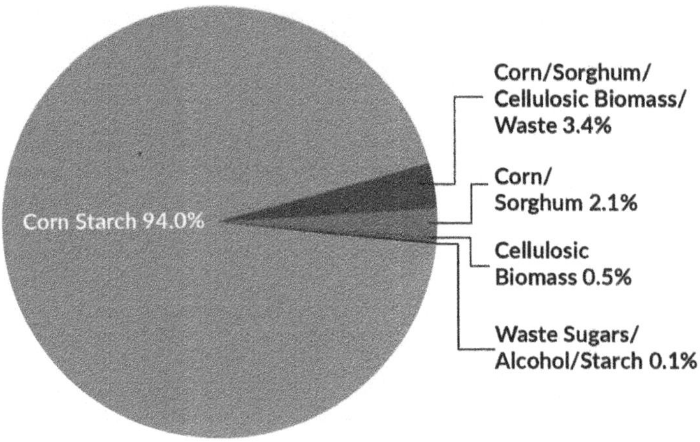

Fig. 1.3 US ethanol production by feedstock type. reproduced with permission https://ethanolrfa.org/wp-content/uploads/2020/02/2020-Outlook-Final-for-Website.pdf

Table 1.8 Ethanol yield values from different first, second and third generations feedstocks

Feedstock	Ethanol yield (gal/acre)	Ethanol yield (L/ha)
Switch grass	1,150	10,760
Algae	5,000–15,000	46,760–140,290
Wheat	277	2,590
Corn stover	112–150	1,050–1,400
Corn	370–430	3,460–4,020
Cassava	354	3,310
Sweet sorghum	326–435	3,050–4,070
Sugar beet	536–714	5,010–6,680

Based on Özçimen and İnan (2015)

have a higher energy content than biodiesel crops, making them more proficient sources of fuel" (worldwatch.org; www.greencarcongress.com).

"The IEA's forecast shows biofuels production is expected to increase 15% over the next five years, reaching 165 billion liters (43.59 billion gallons). By 2023, biofuels are expected to account for nearly 90% of the renewables used in transport. Fuel ethanol accounts for two-thirds of biofuel production growth, while biodiesel and hydrotreated vegetable oil (HVO) account for the remainder" (http://ethanolproducer.com/articles/15673/iea-predicts-growth-in-global-ethanol-production-through-2023).

In Table 1.8, ethanol yield values from different feedstocks including first and second generations are given (Özçimen and İnan 2015).

References

ANP (2007) Agência Nacional do Petróleo, Gás Natural e Biocombustíveis Badger PC (2002) Ethanol from cellulose: a general review. In: Janick J, Whipkey A (eds) Trends in new crops and new uses. ASHS Press, Alexandria VA USA

Azadi P, Malina R, Barrett SRH, Kraft M (2017) The evolution of the biofuel science. Renew Sustain Energy Rev 2017(76):1479–1484

Badger PC (2002) Ethanol from cellulose: a general review. In: Janick J, Whipkey A (eds) Trends in new crops and new uses. ASHS Press, Alexandria VA USA

Baeyens J, Kang Q, Appels L, Dewil R, Lv Y, Tan T (2015) Challenges and opportunities in improving the production of bio-ethanol. Prog Energy Combust Sci 47:60–88

Bajpai P (2007) Bioethanol. PIRA Technology Report, Smithers PIRA, UK

Balat M (2011) Production of bioethanol from lignocellulosic materials via the biochemical pathway: a review. Energy Convers Manag 52:858–875

Bastos RG (2018) Biofuels from microalgae: Bioethanol. In:Jacob-Lopes E, Zepka LQ, Queiroz MI (eds) Energy from Microalgae, 1st edn. Springer International Publishing AG, Cham, Switzerland, p 229. ISBN 978-3-319-69092-6

Berg C (2004). World Ethanol production. The distillery and bioethanol network. http://www.distill.com/worldethanolproduction.htm

Berg C (1999) World ethanol production and trade to 2000 and beyond. Winchester, VA, USA: The Online Distillery Network for Distilleries and Fuel Ethanol Plants Worldwide. http://www.distill.com/berg/

BP (2006) The global ethanol industry is going through a period of rapid growth. http://www.bp.com/

Branco RHR, Serafim LS, Xavier AMRB (2019) Second generation bioethanol production: on the use of pulp and paper industry wastes as feedstock: review. Fermentation 5:4

Bušic A, Marđetko N, Kundas S, Morzak G, Belskaya H, Ivancic Santek M, Komes S, Novak S, Santek B (2018) Bioethanol production from renewable raw materials and its separation and purification: a review. Food Technol Biotechnol 56:289–311

Campbell CJ, Laherrere JH (1998) The end of cheap oil. Scientif Amer 278(3):60–65

Chandel AK, Chan ES, Rudravaram R, Lakshmi M, Rao Venkateswar, Ravindra P (2007) Economics and environmental impact of bioethanol production technologies: an appraisal. Biotechnol Mole Boil Rev 2(1):014–032

Demirbas A (2007) Progress and recent trends in biofuels. Progr Energ Combust Sci 33:1–18

Demirbas A (2009) Political, economic and environmental impacts of biofuels: a review. Appl Energy 86:S108–S117

Dien BS, Jung HJG, Vogel KP, Casler MD, Lamb JAFS, Iten L, Mitchell RB, Sarath G (2006) Chemical composition and response to dilute acid pretreatment and enzymatic saccharification of alfalfa, reed canary grass and switch grass. Biomass Bioenergy 30(10):880–891

DOE (Department of Energy) (2007a) Office of Energy Efficiency and Renewable Energy. Washington DC, US. http://www.doe.gov

DOE (Department of Energy) (2007b). Fuel cell overview. US Department of Energy. http://hydrogen.energy.gov/

Du C, Zhao X, Liu D, Lin, CSK, Wilson K, Luque R, Clark J (2016) Introduction: An overview of biofuels and production technologies. In: Lin CSK, Wilson K, Clark J (eds) Handbook of biofuels production processes and technologies, 2nd edn. Elsevier, Amsterdam, The Netherlands, p 3. ISBN 978-0-08-100455-5

Dutta K, Daverey A, Lin JG (2014) Evolution retrospective for alternative fuels: first to fourth generation. Renew Energy 69:114–122

EBIO—European Biofuel Association (2006) Bioethanol fuel in numbers. www.ebio.org

EIA (Energy Information Administration) (1999) Biofuel: better for the environment United States

EIA (Energy Information Administration) (2006) Annual energy outlook 2007. US Department of Energy. www.eia.doe.gov/oiaf/ieo/index.html

Ghosh P, Ghose TK (2003) Bioethanol in India: recent past and immediate future. Adv Biochem Eng Biotechnol 85:1
Goldemberg J, Monaco LC, Macedo IC (1993) The Brazilian fuel-alcohol program. In: Johansson T, Kelly H, Reddy AKN, Williams RH, Burnham L (eds) Renewable energy. Island Press, Washington, Sources for fuels and electricity, pp 841–863
Gray KA, Zhao L, Emptage M (2006) Bioethanol. Curr Opin Chem Biol 10:141–146
Hägerdal BH, Galbe M, Grauslund MFG, Lidén G, Zacchi G (2006) Bioethanol: the fuel of tomorrow from the residues of today. Trends Biotechnol 24:549–556
Hamelinck CN, Hooijdonk GV, Faaij APC (2005) Ethanol from lignocellulosic biomass. Biomass Bioenergy 28:384–410
Hazell P, Pachauri RK (2006) Bioenergy and agriculture: promises and challenges. International Food Policy Research Institute
Hazell P, von Braun J (2006) Biofuels: a win–win approach that can serve the poor. International Food Policy Research Institute
Herrera S (2004) Industrial biotechnology—a chance at redemption. Nature Biotechnol 22:671–675
IPCC (2001) Intergovernmental panel on climate change. Climate change: the scientific basis. Cambridge University Press
http://www.distill.com/worldethanolproduction.htm
Hussain A, Arif SM, Aslam M (2017) Emerging renewable and sustainable energy technologies: state of the art. Renew Sustain Energy Rev 71:12–28
Jessel Al (2006) Chevron Products Company 2006 Management Briefing Seminars Traverse City, MI. www.cargroup.org/mbs2006/documents/JESSEL.pdf
Joseph Jr H (2005). Long term experience from dedicated & flex fuel ethanol vehicles in Brazil. Clean Vehicles and Fuels Symposium. Stockholm
Kang Q, Appels L, Baeyens J, Dewil R, Tan TW (2014) Energy-efficient production of cas-sava-based bio-ethanol. Adv Biosci Biotechnol 5:925–939. https://doi.org/10.4236/abb.2014.512107
Kheshgi HS, Prince RC, Marland G (2000) The potential of biomass fuels in the context of global climate change. focus on transportation fuels. Annu Rev Energy Environ 25:199–244
Launder K (1999) Opportunities and constraints for Ethanol-based transportation fuels. lansing: state of Michigan, Department of Consumer & Industry Services, Biomass energy program. http://www.michigan.gov/cis/0,1607,7-154-25676_25753_30083-141676–,00.html
Liew WH, Hassim MH, Ng DKS (2014) Review of evolution, technology and sustainability assessments of biofuel production. J Clean Prod 2014(71):11–29
Luhnow D, Samor G (2006) As Brazil fills up on Ethanol, It weans off, energy imports. Wall Street J Lynch DJ (2006) Brazil hopes to build on its Ethanol success. USA Today
Lynch DJ (2006) Brazil hopes to build on its Ethanol success. USA Today
Lynd LR (1996) Overview and evaluation of fuel ethanol from cellulosic biomass: technology, economics, the environment, and policy. Annu Rev Energy Environ 21:403–465
Lynd LR, Wang MQ (2004) A product-nonspecific frame work for evaluating the potential of biomass-based products to displace fossil fuels. J Ind Ecol 7:17–32
Maity SK (2015) Opportunities, recent trends and challenges of integrated biorefinery: part II. Renew Sustain Energy Rev 2015(43):1446–1466
Manochio C, Andrade BR, Rodriguez RP, Moraes BS (2017) Ethanol from biomass: a comparative overview. Renew Sustain Energy Rev 2017(80):743–755
McMillan JD (1997) Bioethanol production: status and prospects. Renew Energy 10:295–302
Moreira JR, Walter A (2005) Overview on bioenergy activity for transport in Brazil. Presentation at 14th European biomass conference and exhibition
Nastari P (2005) Etanol de Cana-de-Açúcar: o Combustível de Hoje. Presentation at Proalcool—30 anos depois. São Paulo
Nastari P, Macedo IC, Szwarc A (2005) Observations on the Draft Document entitled "Potential for biofuels for transport in developing countries". Presented at the World Bank, Washington
Nigam PS, Singh A (2011) Production of liquid biofuels from renewable resources. Prog Energy Combust Sci 2011(37):52–68

OECD/Food and Agriculture Organization of the United Na-tions (OECD-FAO) Agricultural Outlook (2015). OECD Publishing, Paris, France. https://doi.org/1787/agr_outlook-2015-en

Özçimen D, İnan B (2015). An overview of bioethanol production from algae. biofuels-Status Perspec. Krzysztof Biernat, IntechOpen https://doi.org/10.5772/59305. https://www.intechopen.com/books/biofuels-status-and-perspective/an-overview-of-bioethanol-production-from-algae

Paris Morris D (1993) Ethanol: a 150 year struggle toward a renewable future. Institute for Local Self-Reliance, Washington. www.eere.energy.gov/afdc/pdfs/1854.pdf

Paszner L (2006). Bioethanol: fuel of the future. Pulp Pap Can 107(4):26–27, 29

REN21 (2006). Renewables—Global Status Report. Renewable Energy Policy Network for the 21st Century. www.ren21.net

RFA-Renewable Fuels Association (2019) Ethanol industry outlook. https://ethanolrfa.org/wp-content/uploads/2019/02/RFA2019Outlook.pdf

Renewable Fuels Association (2017). http://www.ethanolrfa.org/2017/02/rfa-releases-2017-ethanol-industry-outlook-pocket-guide/

Renewables (2017). Global status report. REN21 Secretariat, Paris, France, 2017; pp 30, 48. ISBN 978-3-9818107-6-9

RFA—Renewable Fuels Association (2006) Ethanol industry outlook: from niche to nation. www.ethanolrfa.org/objects/pdf/outlook/outlook_2006.pdf

Robak K, Balcerek M (2018) Review of second generation bioethanol production from residual biomass. Food Technol Biotechnol 2018(56):174–187

Rosillo-Calle F, Walter A (2006) Global market for bioethanol: historical trends and future prospects. Energy Sustain Dev X(1):18–30

Sarris D, Papanikolaou S (2016) Biotechnological production of ethanol: biochemistry, processes and technologies. Eng Life Sci 16:307–329

Sebayang AH, Masjuki HH, Ong HC, Dharma S, Silitonga AS, Mahlia TMI, Aditiya HB (2016) A perspective on bioethanol production from biomass as alternative fuel for spark ignition engine. RSC Adv 6:14964–14992

Senthilkimar V, Gunasekaran P (2005) Bioethanol production from cellulosic substrates: engineered bacteria and process integration challenges. J Sci Ind Res (India) 2005(64):845–853

Sims REH, Mabee W, Saddler JN, Taylor M (2010). An overview of second generation biofuel technologies. Bioresour Technol. 2010;101(6):1570–80. https://doi.org/10.1016/j.biortech.2009.11.046

Sun Y, Cheng J (2004) Hydrolysis of lignocellulosic materials for ethanol production: a review. Biores Technol 83:1–11

Tanaka L (2006) Ethanol fermentation from biomass resources: current state and prospects. Appl Microbiol Biotechnol 69:627–642

Turkenburg WC (2000) Renewable energy technologies (Chapter 7). In: Goldemberg J et al (eds) World energy assessment report. United Nations Development Programme UNDP, New York, NY, USA, pp 135–171

Walker GM (2010) Bioethanol: science and technology of fuel alcohol. Ventus Publishing ApS. ISBN 978-87-7681-681-0

Ward OP, Singh A (2002) Bioethanol technology: developments and perspectives. Adv Appl Microbiol 51

Wyman CE, Hinman ND (1990) Ethanol. Fundamentals of production from renewable feedstocks and use as transportation fuel. Appl Biochem Biotechnol 24/25:735–775

Yacobucci B, Womach J (2003) Fuel Ethanol: background and public policy issues. Library of Congress, Washington DC. http://www.ethanol-gec.org/information/briefing/1.pdf

Chapter 2
Historical Perspectives

Abstract Presently ethanol is the major biofuel used all over the world and its use is increasingly extensive. The worldwide prospects are the expansion of the production and consumption of ethanol. This chapter focuses on historical perspectives of bioethanol.

Keywords Biofuels · Bioethanol · Ethanol fuel · Alternative fuel

Ethanol is currently becoming a topical subject. However, its controversial history in reality dates back to the 1800s. "Ethanol was developed as an alternative fuel before the discovery of petroleum by Edwin Drake in 1859. Prior to this year, the energy crisis revolved around finding a replacement for the diminishing supply of whale oil, which was commonly used as a lamp oil. Other lamp oils derived from vegetables and animals were also used, but whale oil was preferred. By the late 1830s, ethanol blended with turpentine (refined from pine trees) was used to replace the more expensive whale oil" (Kovarik et al. 1998; Songstad et al. 2009).

The Energy Information Agency (2005) has very well described the history of ethanol. In 1826, the first real ICE which ran on ethanol and turpentine was developed by Samuel Morey. Nicolaus Otto invented the modern 4-cycle internal combustion engine, used ethanol for powering an early engine in 1860. In the 1850s, ethanol was also used as a lighting fuel, but its use restrained when it was taxed as liquor for helping to pay for the Civil War. Use of ethanol as a fuel continued after the tax was revoked, and fueled Henry Ford's Model T in 1908. The first ethanol mixed with gasoline to be used as an octane booster took place in the 1920s and 1930s, and was in great demand during World War II due to fuel shortage (Bevil 2008).

With the phasing out of Methyl Tertiary Butyl Ether (MTBE) as an oxygenate and a desire to reduce dependence upon imported oil and increase the use of environment friendly fuels, the demand of ethanol increased radically (Bajpai 2007). In 2005, the first Renewable Fuels Standard became law as part of the energy policy of United States. Four billion gallons of ethanol was produced in 2006 and seven and one-half billion gallons was produced by 2012 (Renewable Fuels Association 2005). Since that time, the Energy Independence and Security Act of 2007 signed by United States

P. Bajpai, *Developments in Bioethanol*, Green Energy and Technology,
https://doi.org/10.1007/978-981-15-8779-5_2

President needed use of renewable fuel for increasing to 36 billion gallons per annum by 2022 (Renewable Fuels Association 2008). "The new Renewable Fuels Standard which currently guides national ethanol policy states that only 15 billion gallons of production should be produced from corn grain (starch) —the remaining 22 billion should be produced from other advanced and cellulosic feedstock sources" (www.ag.ndsu.edu/energy/biofuels/energy-briefs/history-of-ethanol-production-and-policy).

In 1925, Henry Ford had quoted ethyl alcohol, ethanol, as "the fuel of the future." He furthermore stated, "The fuel of the future is going to come from apples, weeds, sawdust—almost anything. There is fuel in every bit of vegetable matter that can be fermented." Today Henry Ford's futuristic vision significance can be easily understood.

In ancient times ethanol was known as an intoxicating drink. It is the same alcohol used in beverage alcohol but meets fuel-grade standards. Ethanol that is to be used as a fuel is "denatured" by adding a small amount of gasoline to it. This makes it unfit for drinking. During the late 1800s, ethanol was used in the United States for lamp fuel and sales exceeded 25 million gallons per year (Morris 1993). At the request of large oil companies, the government placed a tax on ethanol during the Civil War. This tax almost destroyed the ethanol industry. In 1906 the tax was lifted and alcohol fuel did well until competition from oil companies greatly reduced its use. The first large scale use of ethanol as a fuel occurred during the early 1900s when petroleum supplies in Europe were short. In America, Henry Ford's Model T and other early 1920s automobiles were originally designed to run on alcohol fuels. Germany and the United States both relied on ethanol to power vehicles for their armies during World War II. After World War II, oil prices decreased which caused the use of ethanol to decrease as well. The limited use of ethanol continued until the oil crisis in the early 1970s. The use of ethanol as a fuel has grown since the late 1970s. It was first used as a gasoline extender because of oil shortages.

In the past, several ethanol and ethanol–gasoline mixtures were used as automotive fuels. During the middle of 1930s, Alcolene and Agrol (alcohol–gasoline blends) brands were sold (US Patent No. 4378; Hunt 1981; Kovarik et al. 1998; Dimitri and Effland 2007; Guo et al. 2015).

In 1973, the Organization of Petroleum Exporting Countries (OPEC) caused gasoline shortages by increasing prices and blocking shipments of crude oil to the United States. The OPEC action called attention to the fact that the United States was extremely dependent on foreign oil. The focus shifted once again to alternative fuels such as ethanol. At that time gasoline containing ethanol was called "gasohol". Later, when gasoline was more plentiful, ethanol-blended gasoline was introduced to increase the octane rating and the name "gasohol" was dropped in favour of names reflecting the higher octane levels. "E-10 Unleaded" and "super unleaded" are examples of names used today.

Ever since the modest inception of a sizable ethanol industry, technology on the whole has risen, thus developing lower cost methods of producing greater quantities of fuel ethanol which are simultaneously more efficient in their use of fossil fuel inputs. These combined effects have helped production of ethanol fuel rise in United States by more than 225% between 2001 and 2005 (Renewable Fuels Association

2006). Ethanol has also been used outside the United States, most notably in Brazil which started a program of government mandated ethanol production in 1975 and has since encouraged production of flex-fuel vehicles (FFVs) and cars fueled entirely by ethanol (Luhnow and Samor 2006). Due in part to this jump start on ethanol production and its geographic advantage in growing sugar cane (an ideal ethanol feedstock), Brazil is one of the biggest producer of ethanol. Brazil is so efficient that it can produce a gallon of ethanol for about one dollar (Luhnow and Samor 2006). The Brazilian ethanol market, which was once dependent on governmental regulation and subsidies, has blossomed into a system that thrives even without regulation. Fuel ethanol production in the United States caught up to Brazil's for the first time, growing by 15% in 2005, as both remained the dominant producers (REN21 2006). Although there are cultural and institutional differences between United States and Brazil, the general pattern of ethanol production and consumption under a regulatory environment in the United States could closely mirror what has happened in Brazil. Their policy effectiveness can be used as a benchmark for the American market.

Brazil started a ground breaking program for producing alcohol for automobiles since 1927, when it installed the first pump alcohol which continued until the early years of the next decade (Mussatto et al. 2010; Bray et al. 2000; Balat and Balat 2009). "The fuel ethanol market was revived in the 1970s when, for economic reasons as the global oil crisis and problems in the international sugar market due to overproduction, the National Alcohol Program (ProAlcool) was created in Brazil in 1975. This program was based on the utilization of sugarcane use as raw material, and was intended to target the large-scale use of ethanol as a substitute for gasoline. With substantial government intervention to increase the supply and demand for ethanol, Brazil has developed institutional capacities and technologies for the use of renewable energy in large scale. In 1984, most new cars sold in Brazil required hydrated bio-ethanol (96% bio-ethanol + 4% water) as fuel. As the sugar-ethanol industry matured, policies evolved, and the ProÁlcool program was phased out in 1999, permitting more incentives for private investment and reducing government intervention in allocations and pricing. Although Brazilians have driven some cars that run exclusively on ethanol since 1979, the introduction of new engines that let drivers switch between ethanol and gasoline has transformed what was once an economic niche into the planet's leading example of renewable fuels. Widespread availability of flex-fuel vehicles (promoted through tax incentives) combined with rising oil prices have led to rapid growth in bio-ethanol and sugarcane production since 2000. Today, more than 80% of Brazil's current automobile production has flex-fuel capability" (Kline et al. 2008; Mussatto et al. 2010).

In the United States, the combination of increasing taxes, a rigorous campaign by major oil producing countries and availability of inexpensive petrol efficiently extinguished ethanol as a transport fuel in the beginning of the 20th century (Rossillo-Calle and Walter 2006). The wish to encourage the production and use of bioethanol started again in the early of 1980, principally to rejuvenate the farming sector at a time of oversupply of agricultural produce (Johnson and Rosillo-Calle 2007; Goldemberg et al. 2008).

Table 2.1 History of Bioethanol

The first ICE which ran on ethanol and turpentine was developed by Samuel Morey in 1826
In 1860, Nicholas August Otto used blend of ethanol as a fuel in engine
In 1893, Rudolph Diesel imagined potential of pure vegetable oil for powering machines and invented compression-ignited diesel engines
Henry Ford mentioned ethanol as "fuel of the future" and used it in 1906–1925 to power tractors and his model T cars
Alexander Graham Bell emphasized copiousness of potential raw materials for ethanol production in 1917
Increase in ethanol requirement because of raw material rationing during First World War I
Use of methanol in Germany during Second World War
In USA, farm chemurgy movement started, to use ethanol as octane booster in gasoline in 1930s
in 1943, ethanol was used to produce 77% of synthetic rubber in the United States
In USSR, in 1960s–1970s biobutanol was produced by fermentation
1970 s energy crisis - caused by the increase of oil production in major industrial nations (Germany and North America, etc.) and ban from other producers
Initial energy crisis took place when Arab countries dropped oil production by 5%, resulting in an increase in the prices of oil in 1973
The first commercial process for production of biodiesel was developed in 1977
Next energy crisis was experienced when Iranian dissent grew into strike in nation's oil refineries, closing down 5% of the oil exports in 1978
Lead gas phase out in 1980s–2000s
In 1989, rapeseed was used for production of biodiesel at the first biodiesel commercial plant
Ethanol was used to a larger extent, following passage of the Clean Air Act Amendments of 1990, which created mandates for gasoline oxygenates for addressing wintertime carbon monoxide problems and summer time ozone problems in some metropolitan areas
In 1996, waste grease was processed into biodiesel on a large scale
E85 was classified as alternate fuel in 2006 by US Department of Energy
Flexible fuel vehicles designed to run on gasoline or gasoline-ethanol blends of up to 85% ethanol (E85)
In 2008–2011, production of ethanol and biodiesel in United States increased by over 40%
In 2008, RFA fuel standards replaced MTBE with ethanol for oxygenating fuel
Commercial bioethanol production in world increased from 4.5 to 23.4 billion gallons in 2000–2013
In 2013, the first plant producing ethanol from celluose entered into commercial operation

Based on Singh and Walia (2016), Songstad et al. (2009), Russel (2003), Abde (2008), Niphadkar et al. (2017), https://www.ethanolhistory.com/

The United States is the world leader in production and use of fuel ethanol. It rebuilt its ethanol industry more slowly as compared to Brazil (RFA 2010). E85 which is a mixture of 85% bioethanol and 15% gasoline is being used in especially designed vehicles. Government has been encouraging the development of this type of blend and a number of motor vehicle producing companies including Ford, Chrysler, and General Motors, have increased the production of flexible fuel vehicles which are able to use gasoline and ethanol mixtures ranging from pure gasoline all the way up to E85. At present, ethanol is the major biofuel used worldwide and it is being more and more extensively used, the prospects are the increase of the production and utilization of ethanol globally (Bastos, 2007).

Historical perspectives of bioethanol are summarised in Table 2.1.

References

Abede M (2008) History of ethanol. In Ethanol: Salvation or Damnation? University of Nebraska-Lincoln College of Journalism and Mass Communications DEEP Report 2008

Bajpai P (2007) Bioethanol. PIRA Technology Report, Smithers PIRA, UK

Balat M, Balat H (2009) Recent trends in global production and utilization of bio-ethanol fuel. Appl Energy 86:2273–82

Bastos VD (2007) Etanol, alcoolquímica e biorrefinarias. BNDES Setorial 25:5–38

Bray SC, Ferreira ER, Ruas DGG (2000) As políticas da agroindústria canavieira e o proálcool no Brasil. Unesp-Marília-Publicações, Marília

Bevill K (2008) Building the 'Minnesota Model'. Ethanol Producer Magazine, pp 114–120

Dimitri C, Effland A (2007) Fueling the automobile: An economic exploration of early adoption of gasoline over ethanol. J Agric Food Indus Org. 5:1–17

Energy Information Administration (2005). Policies to promote non-hydro renewable energy in the United States and Selected countries. Report Release: February 2005

Goldemberg J, Nigro FEB, Coelho ST (2008) Bioenergia no estado de São Paulo: Situação atual, perspectivas, barreiras e propostas. Imprensa Oficial do Estado de São Paulo, São Paulo

Guo MX, Song WP, Buhain J (2015) Bioenergy and biofuels: history, status, and perspective. Renew Sustain Energy Rev 42:712–725

Hunt DV (1981) The Gasohol handbook. Industrial Press, New York

Johnson FX, Rosillo-Calle F (2007) Biomass, livelihoods and international trade. Stockholm Environment Institute, Stockholm, Sweden

Kline KL, Oladosu GA, Wolfe AK, Perlack RD, Dale VH and McMahon M (2008) Biofuel feedstock assessment for selected countries. Oak Ridge National Laboratory Oak Ridge, Tennessee 37831. http://www.osti.gov/bridge

Kovarik B, Ford H, Kettering C (1998) The "fuel of the future". Automot Hist Rev 32:7–27

Luhnow D and Samor G (2006) As Brazil fills up on Ethanol, It weans off, energy imports. Wall Street J

Morris D (1993) Ethanol: a 150 year struggle toward a renewable future. Institute for Local Self-Reliance, Washington. Available at: www.eere.energy.gov/afdc/pdfs/1854.pdf

Mussatto SI, Dragone G, Guimarães PMR, Silva JPA, Carneiro LM, Roberto IC, Vicente A, Domingues L, Teixeira JA (2010) Technological trends, global market, and challenges of bio-ethanol production. Biotechnol Adv 28:817–30

Niphadkar S, Bagade P, Ahmed S (2017) Bioethanol production: insight into past, present and future perspectives. Biofuels. https://doi.org/10.1080/17597269.2017.1334338

REN21 (2006) Renewables—global status report. Renewable Energy Policy Network for the 21st Century. Available at: www.ren21.net
RFA—Renewable Fuels Association (2005) Ethanol industry outlook: homegrown for the homeland. Available at http://www.ethanolrfa.org/objects/pdf/outlook/outlook_2005.pdf
Renewable Fuels Association (2006) Ethanol industry outlook: from niche to nation. Available at: www.ethanolrfa.org/objects/pdf/outlook/outlook_2006.pdf
Renewable Fuels Association. (2008). Internet site: http://www.ethanolrfa.org/industry/statistics/#C
Renewable Fuels Association (2010) www.ethanolrfa.org/industry/statistics
Rossillo-Calle F, Walter A (2006) Global market for bio-ethanol: historical trends and future prospects. Energ Sustain Dev 10:20–32
Russell LS (2003) A heritage of light: lamps and lighting in the early Canadian home. University of Toronto Press, Toronto
Singh, RS Walia A (2016) Biofuels historical perspectives and public opinions. In Singh RA, Pandey A, Gnansounou E (eds) Biofuels, production and future perspective, 1st ed. CRC Press, pp 3–23
Songstad DD, Lakshmanan P, Chen J, Gibbons W, Hughes S, Nelson R (2009) Historical perspective of biofuels: learning from the past to rediscover the future. Vitro Cell Dev Biol Plant 45:189–192
US Patent No. 4378, Issued April 1, 1826. https://www.ag.ndsu.edu/energy/biofuels/energy-briefs/history-of-ethanol-production-and-policy; https://www.ethanolhistory.com/

Chapter 3
Key Drivers and Trends

Abstract Key drivers and trends in bioethanol are presented in this chapter. The key drivers are energy, environment and economy. Key drivers for ethanol demand, energy security and climate change are considered to be the most important objectives reported by nearly all countries that engage in bioenergy development activities.

Keywords Biofuels · Bioenergy · Bioethanol · Energy · Environment · Economy · Fuel ethanol

When evaluating key drivers for ethanol demand, energy security and climate change are considered to be the most important objectives reported by nearly all countries that engage in bioenergy development activities (Bajpai 2007). The key drivers are presented in Table 3.1.

Throughout the world concern with clean air is a social and political priority; for example, the necessity of reducing pollutant emissions and achieving targets defined by Kyoto Protocol. Increasing dependency on imported energy supply, especially in a context of rising of oil prices, is also a general concern, particularly in the United States and European Union. Another most important driving force for ethanol production is generation of huge amount of new employment. Ethanol industry in Brazil is responsible for about one million direct jobs, approximately 50% in the sugarcane production. Indirect jobs are estimated to be between to 2.5 and 3.0 million. However, it should be mentioned that this high employment is partly due to low level of mechanization of agricultural activities, as well as poor automation at the industrial site (Rosillo-Calle and Walter 2006). World Bank reports that biofuel industries require about 100 times more workers per unit of energy produced than the fossil fuel industry. Transportation, including emissions from the production of transport fuels, is responsible for about one-quarter of energy-related greenhouse gas (GHG) emissions, and that share is rising. A United States Report (https://www.bio.org/EconomicImpactAdvancedBiofuels.pdf) has analyzed how growth of an advanced biofuel industry impacts on job creation, economic output, energy security and investment opportunities. For example, biofuels industry could create 29,000 new jobs and $5.5 billion in economic growth over the next three years and could ultimately create

P. Bajpai, *Developments in Bioethanol*, Green Energy and Technology,
https://doi.org/10.1007/978-981-15-8779-5_3

Table 3.1 Key drivers

Energy – Reduce crude oil imports – Diversify fuel mix – Hi-Octane conserves crude oil
Environment – Cleaner emissions profile; CO, UBH, NOx, particulates – Greenhouse Gas friendly – Lower urban health costs
Economy – Reduces trade deficit (60% of USD 50bn US deficit = foreign oil) – Create new jobs, broader tax base – Create new investment – Supports agriculture with significant new crop use

800,000 new jobs by 2022 with a positive effect on output of $148.7 billion. It is estimated that in the United States, the cumulative total of avoided petroleum imports over the period 2010–2022 would exceed $350 billion. To stimulate the further development of United States bioethanol, regulators should approve the deployment of E15 (15% ethanol, 85% gasoline) and to extend the tax credit for all ethanol feedstocks

USDA released a report on Feb. 11, 2013 detailing its agricultural projections for the next 10-year period. According to the paper, titled "USDA Agricultural Projections to 2022," long-run developments for global agriculture reflect steady world economic growth and continued global demand for biofuels. Combined, the report said these factors support longer-run increases in consumption, trade and the price of agricultural products. Regarding biofuels, the report notes that demand for ethanol and biodiesel feedstocks is projected to continue growing, but at a slower pace than in recent years. The USDA also specified that expansion will continue to depend on biofuel policies. The dominance of the United States, Brazil, European Union member countries, Argentina, Canada, China and Indonesia in biofuel markets is expected to continue over the next 10 years. According to the USDA, these countries accounted for 90% of world biofuel production, consumption and trade in 2012. Over the next decade, aggregate ethanol production in these countries is expected to increase by 40%. Biodiesel production is expected to rise 30%. In the European Union, corn and wheat feedstocks are expected to account for more than 80% of the expansion of corn ethanol. Ethanol production in Brazil is also expected to increase substantially. The USDA projects that sugarcane ethanol production within the country will increase by 90%, primarily to meet increasing domestic demand for transportation fuel with higher ethanol blends. However, exports to the European Union and United States are also expected to increase. Corn-based ethanol production is expected to double in Argentina by 2022. In Canada, ethanol production is expected to increase by 35%, with corn imports accounting for an increasing share of the feedstock. According to the USDA, China used 4.6 million tons of corn and 1 million tons of wheat to produce ethanol in 2012. Due to policies to limit the expansion of grain- and oilseed-based biofuel production, no significant expansion is expected.

Regarding corn production in United States, the USDA projects corn acreage will remain high in the near term, with normal yields leading to an increase in production and recovery of corn use. According to the USDA after several years of adjusting markets, increasing producer returns are expected to lead to gradually increasing corn acreage in a range of 88 million to 92 million after 2015. While projected increases in corn-based ethanol production are expected to be much smaller over the next decade, ethanol will remain a strong presence in the sector. The report notes that approximately 35% of total corn use is expected to go to ethanol production through 2022.

More than 30 countries have introduced or are interested in introducing programmes for fuel ethanol (Rosillo-Calle and Walter 2006). Other countries, but to a lower extent, have done the same regarding biodiesel. Thus, the ethanol experience is so far much more important than with biodiesel, excluding Europe where the prospects for biodiesel use are much better than fuel ethanol due to the availability of the feedstock.

Developing countries have a reasonable good potential for biofuels production due to the availability of land, better weather conditions and the availability of cheaper labour force. Other important issue to be taken into account is that strengthening rural economies is an imperative task for these countries. Obviously each country is case-specific and a careful analysis is required to assess the pros and cons of large-scale biofuels production, particularly competition for land and water for food production and potential pressures on food prices (Hazell and von Braun 2006; Hazell and Pachauri 2006).

From an environmental perspective, first it should be highlighted the benefits of phasing out lead from gasoline, as lead has adverse neurological effects. Hydrated ethanol has higher octane number than regular gasoline (Joseph 2005), and its use in blends allows the phasing out of lead at low cost. This would be a very important advantage of ethanol use in countries where lead is still in use, as is the case of many African and some Asian and Latin American countries.

Changing over towards clean and renewable fuel from crude oil based fuels is the need of developing countries to protect the environment. Large-scale use of biofuels is one of the main strategies for the reduction of GHG emissions (IPCC 2001). Despite the fact developing countries currently do not have biding GHG reduction targets under the Kyoto Protocol, there are two main aspects to be considered. First, under the Clean Development Mechanism (CDM), developing countries can sell credits to those with reduction commitments, as far biofuels programs are not business as usual. Considering a typical Brazilian figure of 2.7 kg of carbon dioxide equivalent avoided per litre of anhydrous ethanol, biofuels use could represent additional income of US$ 0.02–0.05 per litre (on credits in the range US$7–20 per tonne of carbon dioxide equivalent), value that should be compared with production costs in the US$ 0.23–0.28 per litre range (Nastari 2005; Nastari et al. 2005). Second, climate change effects are supposed to be worst in developing countries and it is important to start acting.

Table 3.2 Estimated GHG reductions for different feedstock

Feedstock	GHG reduction (%)
Fibers (switchgrass, poplar)	70–110
Wastes (waste oil, harvest residues, sewage)	65–100
Sugars (sugar cane, sugar beet)	40–90
Vegetable oils (rapeseed, sunflower seed, soybeans)	45–75
Starches (corn, wheat)	15–40

Based on www.worldwatch.org/biofuels-transportation-selected-trends-and-facts

Full development of the international market in fuel ethanol will require the diversification of production, in terms of both feedstock's and number of producing countries, technological development in the manufacturing field, favourable policies to induce market competitiveness and sustainable development. Bioethanol production based on lignocellulosic biomass, is the technology of the future. Lignocellulosic ethanol is made from a wide variety of plant materials, including wood wastes, crop residues and grasses, some of which can be grown on marginal lands not suitable for food production (Ghosh and Ghose 2003). Lignocellulosic raw materials minimize the potential conflict between land use for food (and feed) production and energy feedstock production. The raw material is less expensive than conventional agricultural feedstock and can be produced with lower input of fertilizers, pesticides, and energy. Biofuels from lignocellulose generate low net greenhouse gas emissions, reducing environmental impacts, particularly climate change (Hägerdal et al. 2006).

The GHG balance of biofuels varies dramatically depending on such factors as feedstock choice, associated land use changes, feedstock production system, and the type of processing energy used. In general, most currently produced biofuels have a solidly **positive GHG balance**. The greatest GHG benefits will be achieved with cellulosic inputs as mentioned above. Energy crops have the potential to **reduce GHG emissions** by more than 100% (relative to petroleum fuels) because such crops can also sequester carbon in the soil as they grow. The estimated GHG reductions for different feedstock is shown in Table 3.2.

References

Bajpai P (2007) Bioethanol. PIRA Technology Report, Smithers PIRA, UK

Ghosh P, Ghose TK (2003) Bioethanol in India: recent past and immediate future. Adv Biochem Eng Biotechnol 85:1

Hägerdal BH, Galbe M, Grauslund MFG, Lidén G, Zacchi G (2006) Bioethanol: the fuel of tomorrow from the residues of today. Trends Biotechnol 24:549–556

Hazell P, Pachauri RK (2006) Bioenergy and agriculture: promises and challenges. International Food Policy Research Institute

Hazell P, von Braun J (2006) Biofuels: a win–win approach that can serve the poor. International Food Policy Research Institute

IPCC (2001) Intergovernmental panel on climate change. Climate change: the scientific basis. Cambridge University Press

Joseph Jr H (2005) Long term experience from dedicated & flex fuel ethanol vehicles in Brazil. In: Clean vehicles and fuels symposium, Stockholm

Nastari P (2005) Etanol de Cana-de-Açúcar: o Combustível de Hoje. Presentation at Proalcool—30 anos depois. São Paulo

Nastari P, Macedo IC, Szwarc A (2005) Observations on the Draft Document entitled "Potential for biofuels for transport in developing countries". Presented at the World Bank, Washington

Rosillo-Calle F, Walter A (2006) Global market for bioethanol: historical trends and future prospects. Energy Sustain Dev **X**(1):18–30

Chapter 4
Chemistry, Types and Sources of Ethanol

Abstract Chemistry, types and sources of ethanol are presented in this chapter. The most important raw materials suitable for use in bioethanol production are feedstocks rich in sugar, starches and cellulosic materials. The production methods vary depending on whether or not the raw material is rich in fiber.

Keywords Biofuels · Bioethanol · Sugar · Starches · Cellulosic materials · Agricultural waste · Forestry waste · Municiple solid waste

4.1 Chemistry of Ethanol

Ethanol is flammable, volatile, and colorless, just like other alcohols. It has a characteristic odour. Ethanol has psychoactive properties. Other types of alcohol also have such properties, but ethyl alcohol is significantly less toxic to humans as compared to methanol or isopropanol. Ethanol is volatile, flammable liquid. It is produced by the fermentation of different sugar, starches, cellulose etc.. Ethanol is also named as ethyl alcohol or grain alcohol. Its chemical formula is C_2H_6O, or can be written as C_2H_5OH or CH_3CH_2OH. It has one methyl (–CH_3) group, one methylene (–CH_2–) group, and one hydroxyl (–OH) group. It has a characteristic, agreeable odor. In dilute aqueous solution, it has a somewhat sweet flavor, but in more concentrated solutions it has a burning taste. Ethanol, is an alcohol, a group of chemical compounds whose molecules contain a hydroxyl group, –OH, bonded to a carbon atom. It may be shown as:

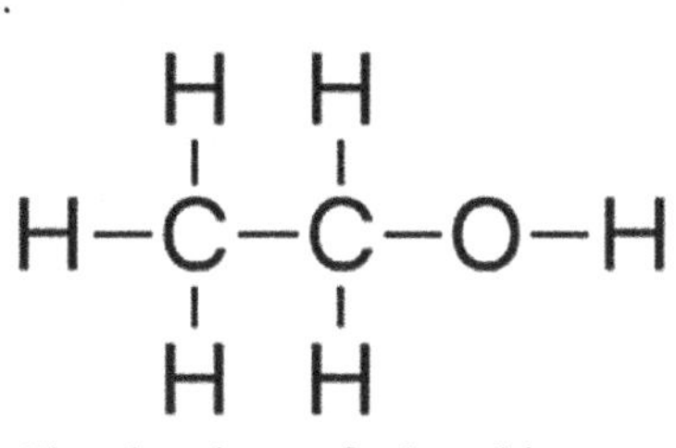

The chemistry of ethanol is presented in Table 4.1.

P. Bajpai, *Developments in Bioethanol*, Green Energy and Technology,
https://doi.org/10.1007/978-981-15-8779-5_4

Table 4.1 The chemistry of ethanol

Acid–base chemistry Ethanol's hydroxyl proton is very weakly acidic; it is an even weaker acid than water. Ethanol can be quantitatively converted to its conjugate base, the ethoxide ion ($CH_3CH_2O^-$), by reaction with an alkali metal such as sodium. This reaction evolves hydrogen gas: $2CH_3CH_2OH + 2Na \rightarrow 2CH_3CH_2ONa + H_2$
Nucleophilic substitution In aprotic solvents, ethanol reacts with hydrogen halides to produce ethyl halides such as ethyl chloride and ethyl bromide via nucleophilic substitution: $CH_3CH_2OH + HCl \rightarrow CH_3CH_2Cl + H_2O$ $CH_3CH_2OH + HBr \rightarrow CH_3CH_2Br + H_2O$ Ethyl halides can also be produced by reacting ethanol by more specialized halogenating agents, such as thionyl chloride for preparing ethyl chloride, or phosphoroustribromide for preparing ethyl bromide
Esterification Under acid-catalysed conditions, ethanol reacts with carboxylic acids to produce ethyl esters and water: $RCOOH + HOCH_2CH_3 \rightarrow RCOOCH_2CH_3 + H_2O$ The reverse reaction, hydrolysis of the resulting ester back to ethanol and the carboxylic acid, limits the extent of reaction, and high yields are unusual unless water can be removed from the reaction mixture as it is formed. Esterification can also be carried out using more a reactive derivative of the carboxylic acid, such as an acyl chloride or acid anhydride Ethanol can also form esters with inorganic acids. Diethyl sulfate and triethyl phosphate, prepared by reacting ethanol with sulfuric and phosphoric acid, respectively, are both useful ethylating agents in organic synthesis. Ethyl nitrite, prepared from the reaction of ethanol with sodium nitrite and sulfuric acid, was formerly a widely-used diuretic
Dehydration Strong acids, such as sulfuric acid, can catalyse ethanol's dehydration to form either diethyl ether or ethylene: $2\ CH_3CH_2OH \rightarrow CH_3CH_2OCH_2CH_3 + H_2O$ $CH_3CH_2OH \rightarrow H_2C = CH_2 + H_2O$ Which product, diethyl ether or ethylene, predominates depends on the precise reaction conditions
Oxidation Ethanol can be oxidized to acetaldehyde, and further oxidized to acetic acid. In the human body, these oxidation reactions are catalysed by enzymes. In the laboratory, aqueous solutions of strong oxidizing agents, such as chromic acid or potassium permanganate, oxidize ethanol to acetic acid, and it is difficult to stop the reaction at acetaldehyde at high yield. Ethanol can be oxidized to acetaldehyde, without overoxidation to acetic acid, by reacting it with pyridinium chromic chloride
Combustion Ethanol combusting in the confines of an evaporating dish Combustion of ethanol forms carbon dioxide and water: $C_2H_5OH + 3\ O_2 \rightarrow 2\ CO_2 + 3\ H_2O$

The word alcohol derives from Arabic al-kuhul, which denotes a fine powder of antimony used as an eye makeup. Alcohol originally referred to any fine powder, but medieval alchemists later applied the term to the refined products of distillation, and this led to the current usage. Ethanol melts at −114.1 °C, boils at 78.5 °C, and has a density of 0.789 g/mL at 20 °C. Its low freezing point has made it useful as the fluid in thermometers for temperatures below −40 °C, the freezing point of mercury, and for other low-temperature purposes, such as for antifreeze in automobile radiators. (Table 4.2). The molecular weight is 46.07. One gallon of 190 proof ethanol weighs 6.8 lb. Ethanol has no basic or acidic properties. When burned, ethanol produces a pale blue flame with no residue and considerable energy, making it an ideal fuel. Ethanol mixes readily with water and with most organic solvents. It is also useful as a solvent and as an ingredient when making many other substances including perfumes, paints, lacquer, and explosives. The *flash point* of ethanol is the lowest temperature (i.e. 12.8 °C) where enough fluid can evaporate to form an ignitable concentration of vapour and characterizes the temperature at which ethanol becomes flammable in air. The ignition point of ethanol is the minimum temperature at which it is able to burn

Table 4.2 Physico-chemical properties of Ethanol

Molecular formula	C_2H_5OH
Molecular mass	46.07 g/mol
Appearance	Colourless liquid (between −117 °C and 78 °C)
Water solubility	Miscible
Density	0.789 kg/l
Boiling temp	78.5 °C (173 °F)
Freezing point	−117 °C
Flash point	12.8 °C (lowest temperature of ignition)
Ignition temp 425 °C	425 °C
Explosion limits	Lower 3.5% v/v; upper 19% v/v
Vapour pressure	@38 °C) 50 mmHg
Higher heating value (at 20 °C)	29,800 kJ/kg
Lower heating value (at 20 °C)	21,090 kJ/L
Specific heat, Kcal/Kg	60 °C
Acidity (p*K*a)	15.9
Viscosity	1.200 mPa·s (20 °C)
Refractive index (*n*D)	1.36 (25 °C)
Octane number	99
Carbon (wt)	52.1%
Hydrogen (wt)	13.1%
Oxygen (wt)	34.7%
C/H ratio	4

Based on Walker (2010)

independently (i.e.425 °C). Ethanol has a high octane rating, which is a measure of a fuel's resistance to pre-ignition, meaning that internal combustion engines using ethanol can have a high compression ratio giving a higher power output per cycle. Regular petrol (gasoline) has an average octane rating of 88. Ethanol's higher octane rating increases resistance to engine knocking, but vehicles running on pure ethanol have fuel consumption (miles per gallon or kilometres per litre) 10–20% less than petrol (but with no loss in engine performance/acceleration).

Ethanol has been made since ancient times by the fermentation of sugars. All beverage ethanol and more than half of industrial ethanol is still made by this process. Simple sugars are the raw material. Zymase, an enzyme from yeast, changes the simple sugars into ethanol and carbon dioxide. The fermentation reaction, represented by the simple equation

$$C_6H_{12}O_6 \rightarrow 2\ CH_3CH_2OH + 2\ CO_2$$

It is actually very complex, and impure cultures of yeast produce varying amounts of other substances, including glycerine and various organic acids. In the production of beverages, such as whiskey and brandy, the impurities supply the flavor. Starches from potatoes, corn, wheat, and other plants can also be used in the production of ethanol by fermentation. However, the starches must first be broken down into simple sugars. An enzyme released by germinating barley, diastase, converts starches into sugars. Thus, the germination of barley, called malting, is the first step in brewing beer from starchy plants, such as corn and wheat.

4.2 Types of Ethanol

Ethanol can be produced in two forms—hydrous and anhydrous. Hydrous ethanol is usually produced by distillation from biomass fermentation, and it contains some water residue. It is suitable for use as neat spark ignition fuel in warm climates such as that in Brazil (Bajpai 2007). A further process of dehydration is required to produce anhydrous ethanol (100% ethanol) for blending with petrol. Anhydrous ethanol can be used as an automotive fuel by itself or can be mixed with petrol in various proportions to form a petrol/ethanol blend.Anhydrous ethanol is typically blended up to 10% by volume in petrol, known as E10, for use in unmodified engines. Historically, the United States has supported the use of E10 blends and more recently Europe has adopted E10 blends. Certain materials in vehicles commonly used with petrol fuel are incompatible with alcohols and varying degrees of modification are required depending on the percentage blend of ethanol with petrol. For this reason in the European Union all member states are required to ensure that fuel grade E5 is available in the market as a protection grade for older vehicles that are not compatible to run on E10.

4.3 Raw Materials

One of the great merits of bioethanol consists in the enormous variety of raw materials, from which it can be produced. The production methods vary depending on whether or not the raw material is rich in fiber. The basic materials for producing biofuels must have certain features, including high carbon and hydrogen concentrations and low concentrations of oxygen, nitrogen and other organic components. The following is a brief description of some of the most important raw materials suitable for use in bioethanol production. These can be classified in to three groups (Table 4.3).

(i) Feedstocks rich in sugar
(ii) Starches
(iii) Cellulosic materials.

Agricultural waste available for ethanol conversion includes crop residues such as wheat straw, corn stover (leaves, stalks, and cobs), rice straw, and bagasse (sugar cane waste).

Forestry waste includes underutilized wood and logging residues; rough, rotten, and salvable dead wood; and excess saplings and small trees (Walker 2010).

Municiple solid waste contains some cellulosic materials, such as paper, Energy crops, developed and grown specifically for fuel, include fast-growing trees, shrubs, and grasses such as hybrid poplars, willows, Reed canary grass, Alfalfa *Miscanthus x giganteus* (hybrid of *M. sinensis* and *M. sacchariflorus*) and switchgrass.Switchgrass (Panicum virgatumn) is one source likely to be tapped for ethanol production because of its potential for high fuel yields, hardiness, and ability to be grown in diverse areas. This is a perennial herbaceous plant that grows mainly in the United States. Its ethanol yield per hectare is the same as for wheat. It responds to nitrogen fertilizers and can

Table 4.3 Major resources for bioethanol production

Sugary materials
Sugarcane
Sugarbeet
Sweet sorghum
Cheese whey
Fruits (surplus)
Confectionery industrial waste
Starchy materials
Grains (maize, wheat, triticale)
Root crops (potato, cassava, chicory, artichoke)
Inulin (polyfructan)
Cellulosic materials
Wood
Agricultural residues (straws, stover)
Municipal solid waste
Waste paper, paper pulp

Based on data from Hamelinck et al. (2003), US DOE (2006a)

sequester the carbon in the soil. It is a highly versatile plant, capable of adapting easily to lean soils and marginal farmland (Heaton et al. 2004). Like maize, it is a type C4 plant, i.e. it makes an alternative use of CO_2 fixation (a process forming part of photosynthesis). Most of the genotypes of Panicum virgatum have short underground stems, or rhizomes, that enable them with time to form a grassy carpet. Single hybrids of Panicum virgatum have shown a marked potential for increasing their energy yield (Bouton 2007), but genetic engineering methods on this plant are still in a developmental stage and for the time being only their tetraploid and octaploid forms are known; we also now know that similar cell types (isotypes) reproduce easily.

Trials show current average yields to be about five dry tons per acre; however, crop experts say that progressively applied breeding techniques could more than double that yield. Switchgrass long root system—actually a fifty-fifty split above ground and below—helps keep carbon in the ground, improving soil quality. It is drought-tolerant, grows well even on marginal land, and doesn't require heavy fertilizing. Other varieties including big blue stem and Indian grass are also possible cellulose sources for ethanol production. Researchers estimate that ethanol yield from switch-grass is in the range of 60–140 gal/ton; some say 80–90 gal/ton is a typical figure. It is estimated that the energy output/energy input ratio for fuel ethanol made from switchgrass, is about 4.4 (Iowa State University 2006). The United States Department of Agriculture estimates that by 2030 approximately 129 million acres of excess cropland could be used for energy crops. If 40 million of these acres were utilized for energy crops for biofuels such as ethanol, it would provide a transportation fuel equivalent to 550 million barrels of oil per year.Sugarcane bagasse, the residue generated during the milling process, is another potential feedstock for cellulosic ethanol. Research shows that one ton of sugarcane bagasse can generate 112 gal of ethanol.

Reed canary grass (*Phalaris arundinacea L*) is a perennial grass that grows widely. Its stem components (dry wt) comprise: hexoses (38–45%); pentoses (22–25%); lignin (18–21%). Alfalfa (*Medicago sativa L*) comprises mainly cellulose, hemicellulose, lignin, pectin and protein. *Miscanthus x giganteus* (hybrid of *M. sinensis* and *M. sacchariflorus*) is a perennial grass with a low need for fertilzers and pesticides with a broad temperature growth range. Previously used as an ornamental landscaping, but now an attractive biomass source for biofuels. For example, potential ethanol from miscanthus is around twice that from corn on an acreage basis.

The most suitable waste for converting into bioethanol is the waste from the fruit and vegetable industries, for instance, cotton fiber, milk whey from cheese-making, the waste products of coffee making, and so on. Generally speaking, such waste contains approximately 45% of cellulose (glucose polymer), which can be simultaneously hydrolyzed and fermented to produce ethanol. SSL (Spent Sulfite Liquor) is a byproduct of bisulfate "pulp" manufacturing that can also be fermented to produce ethanol. Waste varies considerably in content from one area to another, but the majority of the volume generally consists of paper (20–40%), gardening waste (10–20%), plastics, glass, metals and various other materials (Prasad et al. 2007).

Jerusalem artichoke (helianthus tuberosus) tubers are rich in inulin (a fructose polymer), which can be used to obtain a syrup for use both in the foodstuffs industry and in the production of ethanol. It was demonstrated (Curt et al. 2006) that, towards the end of the season, the potential for bioethanol production of the stems of clones is 38% of that of the tubers. The plant grows in summer, reaching its maximum height in July and dying in October.

There are certain other feedstocks which can be used for bioethanol production. Marine macroalgae (seaweeds) demand minimal use of agriculture areas and fresh water for cultivation, and represent an interesting biomass resource for bioethanol (Horn et al. 2000a, b). Their attraction as biomass sources for biofuels stems from the fact that seaweeds have growth rates and primary production rates far in excess than those of terrestrial plants. As low-input, high-yielding biomass, seaweeds may represent an example of *third generation*feedstocks for bioethanol production. Brown seaweeds (Phaeophyta) in particular contain storage polysaccharides which are substrates for microbial degradation. They contain high amounts of carbohydrates such as alginic acid (structural) and laminaran and mannitol (storage) that can potentially be fermented to ethanol. Alginate typically makes up 30–40% of the dry weight in giant brown seaweeds (kelp). Laminarin is a linear polysaccharide of 1, 3–β-Dglucopyanose and can relatively easily be hydrolysed to fermentable glucose. Unlike lignocellulosic biomass, they have low levels of lignin and cellulose making them more amenable for bioconversion to energy fuels than terrestrial plants. Fermentations of hydroysates derived from the fast-growing *Macrocystis*spp and *Laminaria* spp. hold the greatest potential for marine macroalgal bioethanol. Another substance present in great quantities in the sea is chitin which is a polysaccharide consisting of N acetyl glucosamine monomers. Chitin is a very hard, semi-transparent substance found naturally in the exoskeletons of crabs, lobsters and shrimps. Its structure resembles that of cellulose except one hydroxyl group is replaced by an acetyl amine group. It has been described as "cellulose of the sea" and has potential for bioconversion into chemical commodities, including ethanol. A major co-product of biodiesel production is glycerol, which has the potential to be converted to ethanol by certain bacteria (Dharmadi et al. 2006) and yeasts such as *Candida magnoliae, Zygosaccharomyces rouxii* and *Pachysolen tannophilus.*

Lignocellulose consists of cellulose, hemicellulose and lignin along with smaller amounts of pectin, protein, extractives and ash. The composition of these constituents can vary from one plant species to another. Hardwoods for example have greater amounts of cellulose, whereas wheat straw and leaves have more hemicellulose (Table 4.4). The ratios between various constituents within a single plant vary with age, stage of growth, and other conditions. These polymers are associated with each other in a hetero-matrix to different degrees and varying relative composition depending on the type, species and even source of the biomass (Carere et al. 2008; Chandra et al. 2007; Fengel and Wegener 1984). The relative abundance of cellulose, hemicellulose, and lignin are inter alia, key factors in determining the optimum energy conversion route for each type of lignocellulosic biomass (Mckendry 2002). Lignocellulosic feedstocks need aggressive pretreatment to yield a substrate easily hydrolyzed by commercial cellulolytic enzymes, or by enzyme producing microorganisms, to liberate sugars for fermentation. Cellulose is the main constituent of plant cell wall conferring structural support and is also present in bacteria, fungi, and

Table 4.4 Cellulose, hemicellulose and lignin contents in lignocellulosic biomass

	Cellulose	Hemicellulose	Lignin
Hardwoods	40–55	24–40	18–25
Softwoods	45–50	25–35	25–35
Wheat straw	30	50	15
Corn cobs	45	35	15
Grasses	25–40	35–50	10–30
Switchgrass	45	31.4	12

algae. When existing as unbranched, homopolymer, cellulose is a polymer of ß-D-glucopyranose moieties linked via ß-(1,4) glycosidic bonds with well documented polymorphs. The degree of polymerization of cellulose chains in nature ranges from 10,000 glucopyranose units in wood to 15,000 in native cotton. The repeating unit of the cellulose chain is the disaccharide cellobiose as oppose to glucose in other glucan polymers (Desvaux 2005; Fengel and Wegener 1984). The cellulose chains (20–300) are grouped together to form microfibrils, which are bundled together to form cellulose fibres. The long-chain cellulose polymers are linked together by hydrogen and van der Waals bonds, which cause the cellulose to be packed into microfibrils. Hemicelluloses and lignin cover the microfibrils. Fermentable D-glucose can be produced from cellulose through the action of either acid or enzymes breaking the beta-(1,4)-glycosidic linkages. In biomass cellulose is present in both crystalline and amorphous forms. Crystalline cellulose contains the major proportion of cellulose, whereas a small percentage of unorganized cellulose chains form amorphous cellulose. Cellulose in its amorphous form is more susceptible to enzymatic degradation. The cellulose microfibrils are mostly independent but the ultrastructure of cellulose is largely due to the presence of covalent bonds, hydrogen bonding and Van der Waals forces. Hydrogen bonding within a cellulose microfibril determines 'straightness' of the chain but interchain hydrogen bonds might introduce order (crystalline) or disorder (amorphous) into the structure of the cellulose (Laureano-Perez et al. 2005).

Hemicellulose is the second most abundant polymer containing about 20–50% of Lignocellulose biomass. It is not chemically homogeneous like cellulose. Hemicellulose has branches with short lateral chains consisting of different types of sugars. These monosaccharides include pentoses (xylose, rhamnose, and arabinose), hexoses (glucose, mannose, and galactose), and uronic acids (4-O-methylglucuronic, D-glucuronic, and D-galactouronic acids). The backbone of hemicellulose is either a homopolymer or a heteropolymer with short branches linked by beta-(1,4)-glycosidic bonds and occasionally beta-(1,3)-glycosidic bonds. Also, hemicelluloses can have some degree of acetylation, for example, in heteroxylan. Hemicelluloses have lower molecular weight compared to cellulose and branches with short lateral chains that are easily hydrolysed (Fengel and Wegener 1984; Saha 2003). Hemicelluloses are found to differ in composition. In agricultural biomass like straw and grasses, hemicelluloses are composed mainly of xylan while softwood hemicelluloses contain mainly. In many plants, xylans are heteropolysaccharides with backbone chains of

1,4-linked-ß-D-xylopyranose units. In addition to xylose, xylan may contain arabinose, glucuronic acid, or its 4-O-methyl ether, acetic acid, ferulic and p-coumaric acids. Xylan can be extracted easily in an acid or alkaline environment while extraction of glucomannan require stronger alkaline environment (Balaban and Ucar 1999; Fengel and Wegener 1984). Among the key components of lignocellulosics, hemicelluloses are the most thermo-chemically sensitive (Hendricks and Zeeman 2009; Levan et al. 1990). Hemicelluloses within plant cell walls are thought to 'coat' cellulose-fibrils and it has been proposed that at least 50% of hemicellulose should be removed to significantly increase cellulose digestibility. Nevertheless, severity parameters should be carefully optimized to avoid the formation of hemicellulose degradation products such as furfurals and hydroxymethyl furfurals which have been reported to inhibit the fermentation process (Palmqvist and Hahn-Hägerdal 2000a, b). For this reason, pretreatment severity conditions are usually a compromise to maximize sugar recovery and depending upon what type of pretreatment method is used hemicellulose could be obtained either as a solid fraction or a combination of both solid and liquid fractions (Chandra et al. 2007).

Lignin is the third most abundant polymer in nature. It is a complex, large molecular structure containing cross-linked polymers of phenolic monomers. It is present in plant cell walls and confers a rigid, impermeable, resistance to microbial attack and oxidative stress. It is present in the primary cell wall and imparts structural support, impermeability, and resistance against microbial attack. Three phenyl propionic alcohols exist as monomers of lignin. These are:

- Coniferyl alcohol (guaiacyl propanol)
- Coumaryl alcohol (p-hydroxyphenyl propanol)
- Sinapyl alcohol (syringyl alcohol).

Alkyl-aryl, alkyl-alkyl, and aryl-aryl ether bonds link these phenolic monomers together. In general, herbaceous plants such as grasses have the lowest contents of lignin, whereas softwoods have the highest lignin contents (Table 4.4) (Hendricks and Zeeman 2009). Lignin is generally accepted as the 'glue' that binds the different components of lignocellulosic biomass together, thus making it insoluble in water. Lignin has been identified as a major deterrent to enzymatic and microbial hydrolysis of lignocellulosic biomass because of its close association with cellulose microfibrils (Avgerinos and Wang 1983). Chang and Holtzapple (2000) showed that biomass digestibility is increased with increasing lignin removal. In addition to being a physical barrier, the harmful effects of lignin include:

- Nonspecific adsorption of hydrolytic enzymes to "sticky" lignin;
- Interference with, and non-productive binding of cellulolytic enzymes to lignin-carbohydrates complexes;
- Toxicity of lignin derivatives to microorganisms.

Different feedstocks contain different amount of lignin that must be removed via pretreatment to enhance biomass digestibility. The lignin is believed to melt during pretreatment and coalesces upon cooling such that its properties are altered; it can subsequently be precipitated (Brownell and Saddler 1987; Converse 1993; Lynd et al. 2002). Delignification (extraction of lignin by chemicals) causes the following effect:

- Biomass swelling
- Disruption of lignin structure
- Increases in internal surface area
- Increased accessibility of cellulolytic enzymes to cellulose fibers.

Although not all pretreatments result in substantial delignification, the structure of lignin may be altered without extraction due to changes in the chemical properties of the lignin. The pretreated biomass becomes more digestible in comparison to the raw biomass even though it may have approximately the same lignin content as non-pretreated biomass.

Cellulose, hemicellulose are composed of chains of sugar molecules. These chains can be hydrolysed to produce monomeric sugars, some of which can be fermented using ordinary baker's yeast.

To attain economical feasibility a high ethanol yield is a necessity. However, producing monomer sugars from cellulose and hemicellulose at high yields is far more difficult than deriving sugars from sugar- or starch-containing crops, e.g. sugar cane or maize. Therefore, although the cost of lignocellulosic biomass is far lower than that of sugar and starch crops, the cost of obtaining sugars from such materials for fermentation into bioethanol has historically been far too high to attract industrial interest. For this reason, it is crucial to solve the problems involved in the conversion of lignocellulose to sugar and further to ethanol. However, the heterogeneity in feedstock and the influence of different process conditions on microorganisms and enzymes makes the biomass-to-ethanol process complex.

Table 4.5 shows research challenges in the field of bioethanol production based on lignocellulosic biomass.

After the cellulose and hemicellulose have been saccharified, the remainder of the ethanol production process is similar to grain-ethanol. However, the different sugars require different enzymes for fermentation.

Table 4.5 Research challenges in the field of bioethanol production based on lignocellulosic biomass

Major challenges
Improving the enzymatic hydrolysis with efficient enzymes, reduced enzyme production cost and novel technology for high solids handling
Developing robust fermenting organisms, which are more tolerant to inhibitors and ferment all sugars in the raw material in concentrated hydrolysates at high productivity and with high concentration of ethanol
Extending process integration to reduce the number of process steps and the energy demand and to re-use process streams for eliminating use of fresh water and to reduce the amount of waste streams

Based on Hahn Hagerdal et al. (2006)

Table 4.6 Biochemical composition of some Lignocellulosic feedstock (% dry basis)

Feedstock	Hardwood (Eucalyptus)	Softwood (Pine)	Grass (Switch grass)
Cellulose	49.50	44.55	31.98
Hemicellulose	13.07	21.90	25.19
Lignin	27.71	27.67	18.13
Ash	1.26	0.32	5.95
Acids	4.19	2.67	1.21
Extractives	4.27	2.88	17.54
Heating values ($GJ_{HHV}/tonne_{dry}$)	19.5	19.6	18.6

Sun and Cheng (2002) and Mosier et al. (2005)

Lignocellulosic crops are promising feedstock for ethanol production because of high yields, low costs, good suitability for low quality land and low environmental impact. Most ethanol conversion systems encountered in literature, have been based on a single feedstock. But considering the hydrolysis fermentation process, it is possible to use multiple feedstock types. Table 4.6 presents biochemical compositions for several suitable feedstock — Eucalyptus, Pinus and Switch grass. Pine has the highest combined sugar content, implying the highest potential ethanol production. The lignin content for most feedstock is about 27% but grasses contain significantly less, and will probably co-produce less electricity.

Cellulosic resources are in general very widespread and abundant. For example, forests comprise about 80% of the world's biomass. Being abundant and outside the human food chain makes cellulosic materials relatively inexpensive feedstocks for ethanol production. Brazil uses sugarcane as primary feedstock whereas in United States more than 90% of the ethanol produced comes from corn. Other feedstocks such as beverage waste, brewery waste and cheese whey are also being utilized. In European union, most of the ethanol is produced from sugar beets and wheat. Crops with higher yields of energy, such as switchgrass and sugar cane, are more effective in producing ethanol than corn. Ethanol can also be produced from sweet sorghum, a dryland crop that uses much less water than sugarcane, does not require a tropical climate and produces food and fodder in addition to fuel. The best farm crop for ethanol production is sugar beets, in terms of gallons of fuel per acre, with the lowest water requirements to grow the crop. The beet plant drives a central taproot deep into the soil and the entire beet is underground, minimizing evaporation). One result of increased use of ethanol is increased demand for the feedstocks. Large-scale production of agricultural alcohol may require substantial amounts of cultivable land with fertile soils and water. This may lead to environmental damage such as deforestation or decline of soil fertility due to reduction of organic matter. About 5% (in 2003) of the ethanol produced in the world is actually a petroleum product. It is made by the catalytic hydration of ethylene with sulfuric acid as the catalyst. It can also be obtained via ethylene or acetylene, from calcium carbide, coal, oil gas, and other sources. Two million tons of petroleum-derived ethanol are produced annually.

The principal suppliers are plants in the United States, Europe, and South Africa. Petroleum derived ethanol (synthetic ethanol) is chemically identical to bio-ethanol and can be differentiated only by radiocarbon dating.

References

Avgerinos GC, Wang DIC (1983) Selective delignification for fermentation of enhancement. Biotechnol Bioeng 25:67–83

Bajpai P (2007) Bioethanol. PIRA Technology Report, Smithers PIRA, UK

Balaban M, Ucar G (1999) The effect of the duration of alkali pretreatment on the solubility of polyoses. Turk J Agric For 23:667–671

Bouton JH (2007) Molecular breeding of switchgrass for use as a biofuel crop. Curr Opin Genet Dev 17:553–558

Brownell HH, Saddler JN (1987) Steam pretreatment of lignocellulose materials for enhanced enzymatic hydrolysis. Biotechnol Bioeng 29:228–235

Carere CR, Sparling R, Cicek N, Levin DB (2008) Third generation biofuels via direct cellulose fermentation. Int J Mol Sci 9:1342–1360

Chandra RP, Bura R, Mabee WE, Berlin A, Pan X, Saddler JN (2007) Substrate pretreatment: the key to effective enzymatic hydrolysis of lignocellulosics? Adv Biochem Eng Biotechnol 108:67–93

Chang VS, Holtzapple MT (2000) Fundamental factors affecting biomass enzymatic reactivity. Appl Biochem Biotechnol 84–86:5–37

Converse AO (1993). Substrate factors limiting enzymatic hydrolysis. In: Saddler JN (ed) Bioconversion of forest and agricultural plant residues. CAB International, Wallinfod, Conn, pp 93–106

Curt MD, Aguado P, Sanz M, Sánchez G, Fernández J (2006) Clone precocity and the use of Helianthus tuberosus L. stems for bioethanol. Ind Crops Prod **24**(3):314–320

Department of Energy (2006) Bioethanol feedstocks. https://www1.eere.energy.gov/biomass/abcs_biofuels.html

Desvaux M (2005) Clostridium cellulolyticum: model organism of mesophilic cellulolytic clostridia. FEMS Microbiol Rev 29:741–764

Dharmadi Y, Muraka A, Gonzalez R (2006) Anerobic fermentation of glycerol by Escherichia coli: a new platform for metabolic engineering. Biotechnol Bioeng 94:821

Fengel D, Wegener G (1984) Wood chemistry, ultrastructure, reactions. Berlin New York

Hahn Hagerdal B, Galbe M, Gorwa-Grauslund MF, Liden G, Zacchi G (2006) Bioethanol—the fuel of tomorrow from the residues of today. Trends Biotechnol 24:549–556

Hamelinck CN, Hooijdonk GV, Faaij APC (2003) Prospects for ethanol from lignocellulosic biomass: techno-economic performance as development progresses. Universiteit Utrecht Copernicus Institute, Science Technology Society, NWS-E-2003-55 ISBN 90-393-2583-4. www.senternovem.nl/mmfiles/149043_tcm24-124362.pdf

Heaton E, Voigt T, Long SP (2004) A quantitative review comparing the yields of two candidate C4 perennial biomass crops in relation to N, temperature and water. Biomass Bioenergy 27:21–30

Hendricks AT, Zeeman G (2009) Pretreatments to enhance the digestibility of lignocellulosic biomass. Bioresour Technol 100:10–18

Horn SJ, Aasen IM, Østgaard K (2000a) Production of ethanol from mannitol by Zymobacter palmae. J Ind Microbiol Biotechnol 24:51–57

Horn SJ, Aasen IM, Østgaard K (2000b) Ethanol production from seaweed extract. J Ind Microbiol Biotechnol 25:249–254

Iowa State University (2006) Biomass economics. https://www.public.iastate.edu/~brummer/ag/biomass2.htm

Laureano-Perez L, Teymouri F, Alizadeh H, Dale BE (2005) Understanding factors that limit enzymatic hydrolysis of biomass: characterization of pretreated corn stover. Appl Biochem Biotechnol 121–124:1081–1099

Levan SL, Ross RJ, Winandy JE (1990) Effects of fire retardant chemical bending properties of wood at elevated temperatures. Research paper FPF-RP-498 Madison, WI: USDA, forest service, 24.

Lynd LR, Weimer PJ, van Zyl WH, Pretorius IS (2002) Microbial cellulose utilization: fundamentals and biotechnology. Microbiol Mol Biol Rev 66:506–577

Mckendry P (2002) Energy Production from biomass (part 1) overview of biomass. Bioresour Technol 83:37–46

Mosier N, Wyman C, Dale B, Elander R, Lee YY, Holtzapple M, ladisch M, (2005) Features of promising technologies for pretreatment of lignocellulosic biomass. Bioresour Technol 96:673–686

Palmqvist E, Hahn-Hägerdal B (2000a) Fermentation of lignocellulosic hydrolysates I: Inhibition and detoxification. Bioresour Technol 74:17–24

Palmqvist E, Hahn-Hägerdal B (2000b) Fermentation of lignocellulosic hydrolysates II: Inhibitors and mechanisms of inhibition. Bioresour Technol 74:25–33

Prasad S, Singh A, Joshi HC (2007) Ethanol as an alternative fuel from agricultural, industrial and urban residues. Resour Conserv Recycl 50(1):1–39

Saha BC (2003) Hemicellulose bioconversion. J Ind Microbiol Biotechnol 2003(30):279–291

Sun Y, Cheng J (2002) Hydrolysis of lignocellulosic materials for ethanol production: a review. Bioresour Technol 83:1–11

Walker GM (2010) Bioethanol: science and technology of fuel alcohol. Ventus Publishing,London. ApS ISBN 978-87-7681-681-0

Chapter 5
Production of Bioethanol

Abstract Bioethanol production processes are discussed in this chapter. First, second, third and fourth generation process technology are discussed. Processing of sugar cane, maize and lignocellulose feedstocks and important developments in the production of cellulosic ethanol are presented.

Keywords Biofuels · Bioethanol · Cellulosic ethanol · First generation process technology · Second generation process technology · Third generation process technology · Fourth generation process technology

5.1 Background

Procresses for production of bioethanol differ significantly depending on the raw material used. However, some of the major steps in the process remain the same, even though they take place in different conditions of temperature and pressure, and they sometimes involve different microorganisms. These stages include hydrolysis (obtained chemically and enzymatically), fermentation and distillation (Olsson et al. 2005). Today, there are four types of process technology called first, second, third and fourth generation technology (Fig. 5.1).

5.1.1 First Generation Process Technology

First generation process technology produces bioethanol from sugars (a dimer of the monosaccharides glucose and fructose) and starch-rich (polysaccharides of glucose) crops such as grain and corn (Table 5.1). Sugars can be converted to ethanol directly but starches must first be hydrolysed to fermentable sugars by the action of enzymes from malt or molds. The technology is well-known but high prices of the raw material and the ethics about using food products for fuel are two major problems.

P. Bajpai, *Developments in Bioethanol*, Green Energy and Technology,
https://doi.org/10.1007/978-981-15-8779-5_5

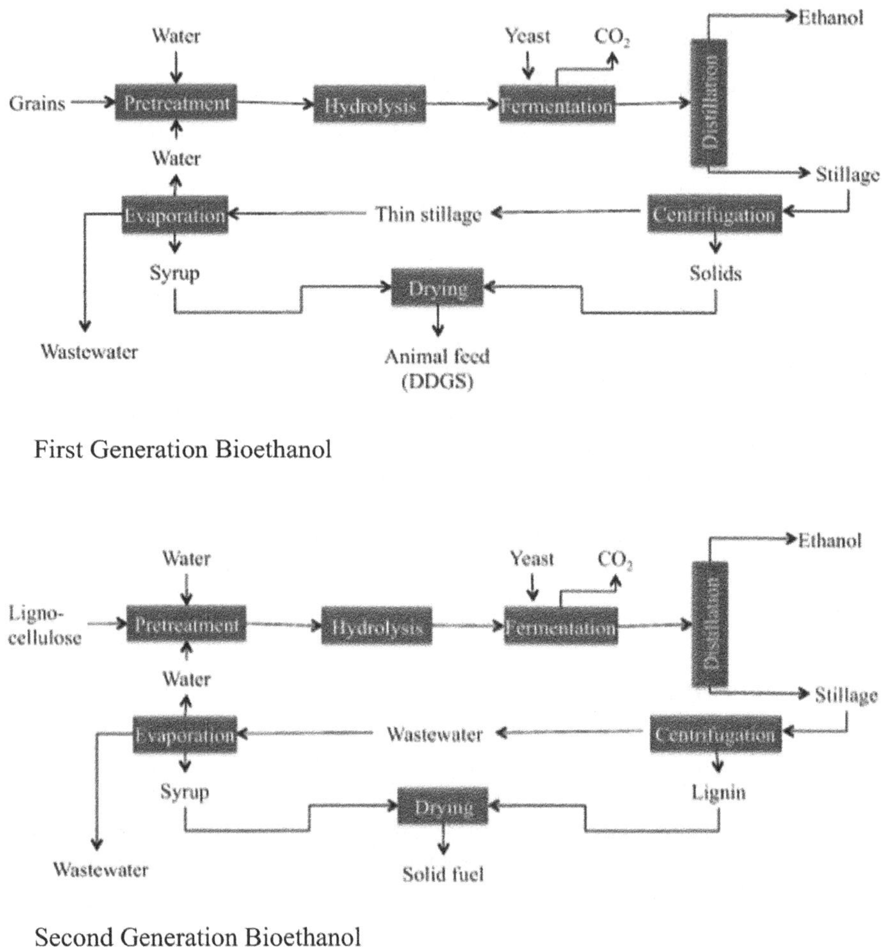

Fig. 5.1 Bioethanol production processes. Lennartsson et al. 2014 (Reproduced with permission)

5.1.2 Second Generation Process Technology

The raw material in second generation is lignocellulosic raw materials (Table 5.1). These are most abundantly available biomass in the world and are found as leaves, peels, bodies, branches, etc. of almost all the existing plants. Therefore, lignocellulosic bioethanol production is certainly a strategy of energy supply, particularly appropriate for countries with agricultural and forestry wastes to be utilized as the input materials. These kinds of materials are cheap but the process technology is more advanced than converting sugar and starch (Fan et al. 1987; Badger 2002; Taherjadeh 1999; US DOE 2010). Basically, the lignocellulosic biomass comprises of lignin, cellulose and hemicelluloses (Figs. 5.2 and 5.3a–c). Cellulose is a linear, crystalline

Table 5.1 First generation and second generation feedstocks for bioethanol

First generation feedstocks
Sugar beet
Sweet sorghum
Sugar cane
Maize
Wheat
Barley
Rye
Grain
Sorghum
Triticale
Cassava
Potato
Second generation feedstocks
Corn stover
Wheat straw
Sugar cane bagasse
Municipal solid waste

Based on Walker (2010)

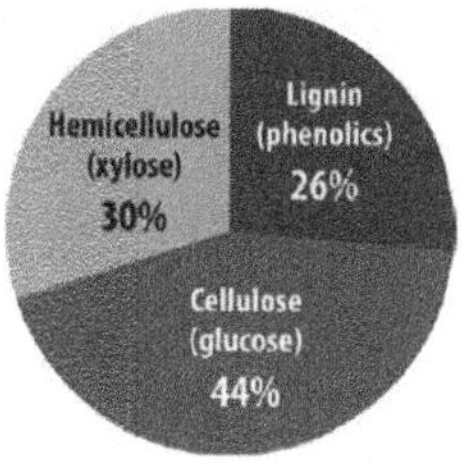

Fig. 5.2 Main constituents of lignocellulosic feedstocks

homopolymer with a repeating unit of glucose strung together beta-glucosidic linkages. The structure is rigid and harsh treatment is required to break it down (Gray et al. 2006). Hemicellulose consists of short, linear and highly branched chains of sugars. In contrast to cellulose, which is a polymer of only glucose, a hemicellulose is a hetero-polymer of D-xylose, D-glucose, D-galactose, D-mannose and L-arabinose. The composition of holocellulose (cellulose + hemicellulose) varies with the origin of the lignocellulosic material.

Burning ethanol obtained from cellulose produces 87% lower emissions than burning petrol, while for the ethanol from cereals the figure is no more than 28%. Ethanol obtained from cellulose contains 16 times the energy needed to produce it, petrol only 5 times and ethanol from maize only 1.3 times. The problem is a matter of how to disrupt the bonds of this molecule in order to convert it into fermentable sugars.

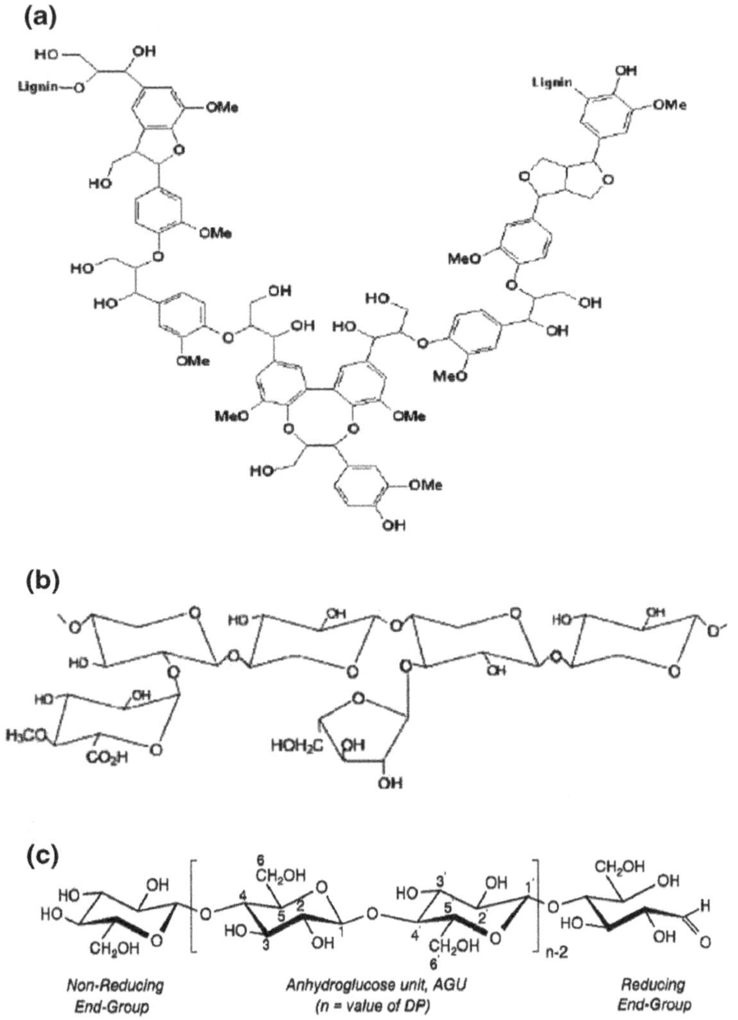

Fig. 5.3 **a** Structure of Lignin (complex crosslinked polymer of aromatic rings (phenolic monomers); very high energy content. Based on Walker (2010). **b** Structure of hemicelluose (Branching polymers of C5, C6, uronic acid, acetyl derivatives). Based on Walker (2010). **c** Structure of Cellulose (composed of D-glucose unite linked by β-1, 4 glycoside bonds). Based on Walker (2010)

In fact, this is unquestionably the type of raw material that is the most complicated to process. Lignin binds together pectin, protein and the two types of polysaccharides, cellulose and hemicellulose, in lignocellulosic biomass. Lignin resists microbial attack and adds strength to the plant. Pretreatment is therefore used to open the biomass by degrading the lignocellulosic structure and releasing the polysaccharides (Fig. 5.4) (Mosier et al. 2005; Ladisch 2003; Hsu et al. 1980). Pretreatment is followed

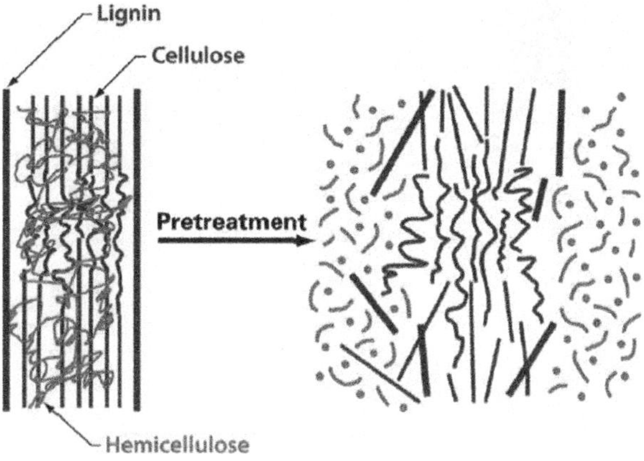

Fig. 5.4 Pre-treatment to open the biomass by degrading the lignocellulosic structure and releasing the polysaccharides. Based on Mosier et al. (2005), Ladisch (2003), Hsu et al. (1980)

by treatment with enzymes which hydrolyse cellulose and hemicellulose respectively. The cellulose fraction releases glucose (C6 monosaccharide—sugar with six carbon atoms) and the hemicellulose fraction releases pentoses (C5 monosaccharide—sugar with five carbon atoms) such as xylose. Out of carbohydrate monomers in lignocellulosic materials, xylose is second most abundant after glucose. Glucose is easily fermented into ethanol, but another fermentation process is required for xylose—for example using special microorganisms. The second generation holds great advantages with the fermentation of biomass in form of agricultural waste materials but there are some challenges such as efficient pre-treatment and fermentation technologies together with environmentally friendly process technology (for example reuse of the process water).

Figure 5.5 shows schematic of a biochemical and Fig. 5.6 shows schematic of thermochemical cellulosic ethanol production process.

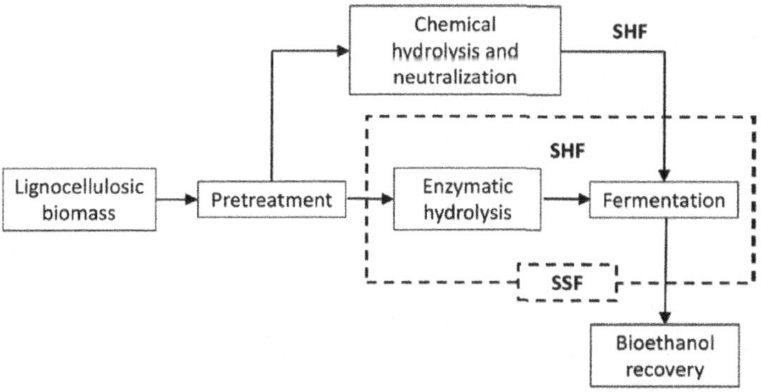

Fig. 5.5 Scheme of lignocellulosic ethanol production. Tran et al. (2019)

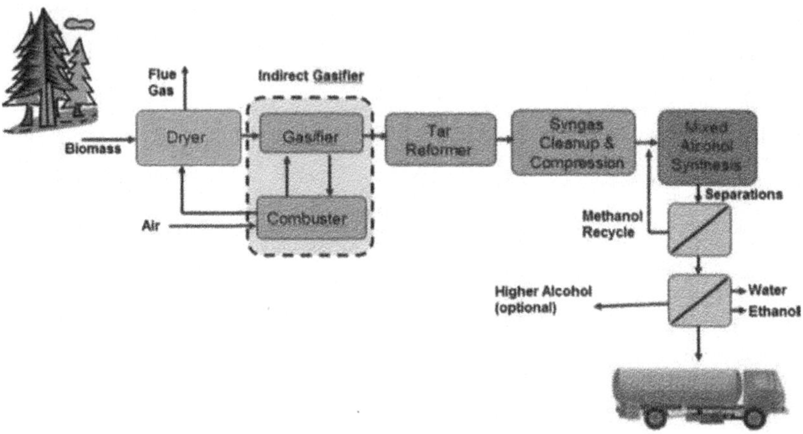

Fig. 5.6 Schematic of a cellulosic ethanol production process through gasification technology. Johnson et al. (2010), Reproduced with permission

5.1.3 Third Generation Process Technology

In the recent years, third-generation bioethanol from microalgae and macroalgae is generating significant interest as well (Doan et al. 2012; https://www1.eere.energy.gov/bioenergy/pdfs/algal_biofuels_roadmap.pdf). Algae are being extensively studied as an alternative substrate of biofuel because of their higher photosynthesis and faster growth rate as compared to the terrestrial plants. The algae are low priced, high-energy, and entirely renewable raw material. It is anticipated that algae will produce more energy per acre in comparison to conventional crops. Algae is able to grow using land and water not fit for food production, thus reducing the pressure on already exhausted water sources. Another benefit of biofuels based on algae is that several types of fuels like diesel, petrol, and jet fuel can be produced.

The energy production from algae dates back to the late 1950s, but these days extensive work is being conducted (Chen et al. 2009). Algae are gigantic group containing several thousands of diverse species which tolerate the option of desired species in accordance with the working environment. They are mostly present in freshwater, marine and terrestrial ecologies. Most of the algal species do not mostly need fresh water and are able to grow under severe situations, like warm or cold deserts, salty habitats, acid waters containing high percentage of heavy metals, sea deep waters and hydrothermal outlets (De Wever et al. 2009; Schmidt et al. 2011; Vinogradova and Darienko 2008; Zettler et al. 2002; Zechman et al. 2010). Certain type of green algae such as Trentepohliales are entirely terrestial and are not found in marine surroundings (Lopez Bautista et al. 2006). This shows that algae can grow in a different environment. Thus, the expoitation of algae as a renewable energy source for production of biofuel is expected to guarantee self adequacy and energy security.

5.1.4 Fourth-Generation Process Technology

"Fourth-generation biofuels can be produced from specially engineered plants or biomass that have higher energy yields or lower barriers to cellulosic breakdown or are able to be grown on nonagricultural land or bodies of water. These biofuels are envisioned as sustainable fuels; obtain higher energy efficiency and environmenta performance. This group includes biofuels which can be made using non-arable land. These do not need destroying of biomass to be converted to fuel. The aim of this technology is to directly convert available solar energy to fuel using inexhaustible, inexpensive and widely available resources. Photobiological solar fuels and electro-fuels are the most advanced biofuels which are presently being researched" (Aro 2016; Bajpai 2019).

According to Aro (2016), "the fourth generation biofuels photobiological solar fuels and electrofuels are expected to bring major breakthroughs in the field of biofuels".

Technology for production of such solar biofuels is an emerging area and is based on direct conversion of solar energy into fuel using raw materials that are inexhaustible, cheap and widely available. This is expected to take place via revolutionary development of synthetic biology as an enabling technology for such a change.

The synthetic biology field is still in its begining and only a few truly synthetic examples have been published so far. For successful progress, there is a need to discover new-to-nature solutions and construct synthetic living factories and designer microbes for effective and direct conversion of solar energy to fuel. In the same way, a combination of photovoltaics or inorganic water splitting catalysts with metabolically engineered microbial fuel production pathways (electrobiofuels) is a powerful emerging technology for efficient production and storage of liquid fuels" (Cameron et al. 2014).

For the production of fourth generation biofuels, raw materials which are unlimited, low priced, and available in large quantitities. Splitting of water by photosynthesis into its constituents by solar energy may become a significant contributor to fuel production worldwide, by artificial photosynthesis and also by direct solar biofuel production techniques (Inganas and Sundstrom 2016). The production of hydrogen as well as the production of reduced carbon based biofuels is possible by simultaneous improved fixation of atmospheric carbon dioxide and ground-breaking design of synthetic metabolic pathways for biofuel production. The production of "designer bacteria" with new functional properties requires groundbreaking scientific advances in several areas of basic research (https://www.link.springer.com).

"Fourth generation biofuels take benefit of synthetic biology of algae and cyanobacteria. This is strongly developing research area. Synthetic biology involves the design and construction of new biological parts, devices and systems, and the redesign of existing, natural biological systems for useful applications. It is becoming possible to design a photosynthetic/non-photosynthetic chassis, either natural or synthetic, for production of biofuels of high quality with high photon-to-fuel conversion efficiency. For the first, second and third generation biofuels, the raw material is either biomass or waste, both being results of yesterday's photosynthesis. Whereas these biofuels often are very useful in a certain region, they are always limited by the availability of the corresponding biomass, which limits their use worldwide" (Aro 2016; Berla et al. 2013; Hays and Ducat 2015; Scaife et al. 2015).

Following methods are used for the production of fourth generation biofuels (Aro 2016)

- By designer photosynthetic microorganisms to produce photobiological solar fuels.
- Combining photovoltaics and microbial fuel production (electrobiofuels).
- Synthetic cell factories or synthetic organelles specifically tailored for production of desired biofuels and high value chemicals—production of which is presently based on fossil fuels.

Newer, innovative methods such as importing particular genes into *Escherchia coli* for breaking down cellulosic biomass, thus producing abundant and inexpensive sugar supply, have a good chance of success, especially in this case because this microorganism is very well studied and very well takes up genetic changes (Niphadkar et al. 2017). Futhermore, *Escherchia coli* is able to grow three times faster as compared to yeast and 100 times faster in comparison to nearly all agricultural microorganisms. The main problem with *Escherchia coli* is that the highest sugar is only 10%, and unless a yield of about 70–90% is achieved, commercialization will not be easy. This problem can be overcome by using genetic engineering methods for increasing the sugar yield, as the metabolic pathway is well known. With fourth generation production systems, biomass crops are being maneuvered to act as effective 'carbon capturing' devices that capture carbon dioxide from the atmosphere and store it in their branches, steps and leaves. This biomass rich in carbon is then converted into biofuels. The important discoveries by scientists in designing trees which store substantially more carbon dioxide in comparison to their ordinary counterparts have opened new propects for the availability of low priced fermentable sugars. This accomplishment was obtained for eucalyptus found in northeastern Asia and for Dahurian Larch found in Siberia, (http://global.mongabay.com/news/bioenergy/2007/10/quick-look-at-fourth-generation.html).

5.2 Feedstock Processing

Processing of bioethanol feedstocks include preprocessing, pretreatment; hydrolysis, and control of microbial contamination. The processing sugar cane, maize and lignocellulose feedstocks are presented below.

5.2.1 Processing of Sugar Crops

The most important sugar producing plants in the world are sugar cane and beet. Two-third of the sugar production in the world are from sugar cane and about one-third is from sugar beet (Bušic et al. 2018). These material can be easily hydrolysed by the invertase enzyme, which is produced by many *Saccharomyces* species. Because of this reason, the pretreatment is not needed for bioethanol production from the sugar rich feedstocks, which makes this bioprocess more feasible as compared to feedstocks containing starch (Linoj et al. 2006). Sugar crops require only a milling process for the extracting sugars, and in this case ethanol could be produced directly from juice or molasses (İçöz et al. 2009).

Use of sugar cane as a feedstock for production of bioethanol provides certain benefits, as it is a semi-perennial crop that does not need several agricultural operations which are generally required for processing of raw crop. Its biomass is utilized for heat and electricity. Sugar cane is less costly than other feedstock used for bioethanol production because these can be processed easily and yield higher productivity (Tabak 2009). Nevertheless, several attempts are being made still aim at improving the production of bioethanol from sugarcane. This involves development of new varieties of sugar cane with higher sugar content and tolerance to diseases, higher yield per hectare and better permanence (Mussatto et al. 2010).

In European countries, sugar beet is used for sugar production. Raw, thin and thick juice, as intermediate produced during sugar beet processing, and also crystal sugar of higher purity, could be converted into bioethanol and or biobased products. Raw sugar beet cossettes are also found appropriate for bioethanol production (Pavlečić et al. 2010, 2017).

"The use of sugar processing intermediates determines bioprocess configuration, their microbiological stability and transport properties. Sugar syrup and granulated sugar are found to serve as substrates for bioethanol production during the entire year. Moreover, they can be used as precursors for diverse chemical intermediates or final products (*example* surfactants)" (hrcak.srce.hr).

Molasses is an important byproduct of the sugar industry. It is used as a feedstock for production of yeast, bioethanol and chemicals. It is also found suitable for production of feedstuff (Roadmap 2012). Total sugars in molasses is in the range of 50–60% (*m/V*), of which ~60% is sucrose, which makes this feedstock suitable for production of bioethanol on large scale (Senthilkumar and Gunasekaran 2008).

Sugar cane and beet molasses are the by-products produced during manufacturing or refining of sucrose from beet and sugar cane. In sugarcane molasses about 46% total sugars are present whereas in sugar beet molasses about 48% of total sugars are present. Molasses is also produced as a by-product during the production of dried citrus pulp containing about 45% sugars. Production of glucose from starch such as corn or grain sorghum also produces molasses. Enzymes or acids are used for starch hydrolysis (Senthilkumar and Gunasekaran 2008).

Whey, a byproduct of cheese manufacture, can be utilized for bioethanol production. It contains around 4.9% (*m/V*) lactose (Ling 2008). Production of ethanol from cheese whey has been studied for converting a waste product into an important energy source (Zafar and Owais 2006; Sileveria et al. 2005). The lactose present in the whey available for ethanol production is more than 4 million tons per year yielding about 2.3 million m3 of ethanol. Because of the very low amount of sugar, a bioethanol plant of modest size would need a large whey volume. The practibility of a new bioethanol plant depends on the cost of whey permeate as substrate and also the price of bioethanol which strongly relates to the production technology and performance of the process (Ling 2008).

Sugar cane contains ~15% sucrose. The juice is pressed from the cane and is fermented by yeast. The juice can be processed either into crystalline sugar or directly fermented to ethanol, as per many industrial plants in Brazil (Fig. 5.7). For sugar production, the juice is clarified with lime and evaporated to form crystals that are centrifuged, leaving a syrupy brown liquid by-product known as molasses. Molasses represents an almost complete fermentation medium as it comprises sugars (sucrose, glucose, fructose), minerals, vitamins, fatty acids, organic acids etc. Additional nitrogen in the form of di-ammonium phosphate is commonly added. The more sucrose from sugarcane stalks that is removed for crystalline sugar production, the poorer the quality of molasses and some molasses contains excess levels of salts and inhibitors produced during heat treatments (furfurals, formic acid and browning reaction products). For bioethanol fermentations, molasses is diluted to 20–25% total sugar treated with sulphuric acid and heated to 90 °C for impurity removal prior to cooling, centrifugation, pH adjustment and addition of yeast. Sugar cane juice can

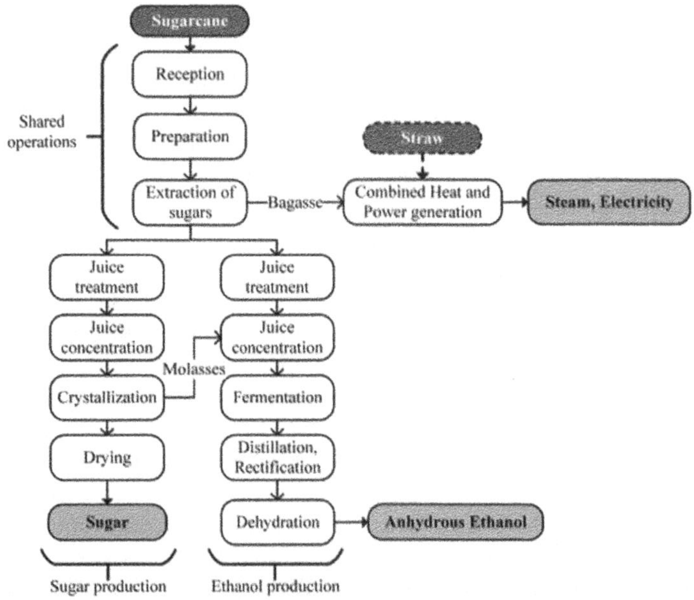

Fig. 5.7 Processing of sugarcane to ethanol. Dias et al. (2015) Reproduced with permission

either be directly fermented, clarified following heat (105 °C) treatment, or mixed with molasses in different proportions. Constituents in molasses that are important for bioethanol production include: sugar content: sugar % (w/w) and degrees Brix, colour, total solids, specific gravity, crude protein, free amino nitrogen, total fat, fibre, minerals, vitamins and substances toxic to yeast. The yeast *S. cerevisiae* is the predominant microorganism employed in industrial molasses fermentations, but another yeast, *Kluyveromyces marxianus* and a bacterium, *Zymomonas mobilis*, have potential in this regard (Bajpai and Margaritis 1983; Senthilkumar and Gunasekaran 2008).

5.2.2 Processing of Cereal Crops

Grain crops such as wheat, corn, barley, or grain sorghum and the root/tubular crops such as potato, sweet potato, Jerusalem artichoke, cassava or arrowroot) have substantial amount of starch (Jobling 2004). Native starch isolated from diverse sources can be utilized for further conversion into bioethanol and other biobased products. The residue from starch isolation consist of proteins and fibre, which shows a great potential for use in food and feed industry (Roadmap 2012). Most of the corn starch is produced in the United States and it represents over 80% of the global market (Jobling 2004). In the United States, corn is a feedstock for producing more than 95% of bioethanol and the remaining is produced from wheat, barley, whey and beverage residues (Solomon et al. 2007). The grain sorghum grown in different areas of the United States shows a growing awareness in producing ethanol from this crop. Moreover, the economic feasibility of bioethanol production from cassava in Thailand was investigated as well (Khanal 2008). In Cassava tubers about 80% starch and 1.5% of proteins are present. Several steps are involved in the pretreatment of cassava tubers for producing ethanol. These steps include: cleaning, peeling, chipping and drying. Afterwards, from these dried cassava chips ethanol is produced (Khanal 2008).

Starch is a mixture containing linear and branched polyglucans. Linear polyglucans are amylose and branched polyglucans are amylopectin. The essential enzyme for starch hydrolysis is α-amylase. It is active on α-1,4 linkage, but not on α-1,6 linkages (Mousdale 2008). For bioethanol production from starchy substrates, it is important to conduct the starch hydrolysis mainly with α-amylases and glucoamylases into glucose syrup, which can be converted into ethanol with *Saccharomyces cerevisiae* yeast. This step is an additional cost in comparison to the ethanol production from feedstocks containing sugar (Soccol et al. 2011).

α-amylases are produced by *Bacillus licheniformis* and genetically modified *Escherichia coli* and *Bacillus subtilis* bacteria, whereas glucoamylases are produced by *Aspergillus niger* and *Rhizopus* sp. fungi (Pandey et al. 2000; Shigechi et al. 2004). Under anaerobic conditions, the yeast *S. cerevisiae* metabolizes converts glucose into ethanol. The maximum theoretical yield of glucose into ethanol is 51% by mass. But, some glucose for cell growth and synthesis of other metabolic products is also used by yeast, thereby reducing the conversion efficiency. In actual case, 40–48% by mass of glucose is converted into ethanol (Lee et al. 2007). When production of ethanol from feedstock containing sugars is compared, ethanol produced from starch improves enzyme application and yeast strains showing higher tolerance of ethanol (Schubert 2006).

Microalgae are sustainable source of biomass for producing biofuel as these are able to convert carbon doixide into lipids and polysaccharides. So, industrial carbon doixide could be collected and used for growing microalgae as part of approach for reducing carbon doixide emission in atmosphere. Microalgae are able to hoard starch as a reserve polysaccharide, which can be utilized for production of third generation bioethanol after the pretreatment. In addition, residual biomass which contains organic matter and minerals after the production of bioethanol can be used as biofertilizer. Therefore, it is apparent that the use of biorefinery concept can significantly improve bioethanol production from microalgae (Li et al. 2014; Tanadul et al. 2014).

The main stages prior to fermentation for processing of starch-based materials are:

- *Cereal cooking*
- *Starch liquefaction*
- *Amylolysis*

In North America, ethanol is produced from corn by using one of two standard processes: wet milling or dry milling (Yacobucci and Womach 2003; Ferguson 2003) (Figs. 5.8 and 5.9). The main difference between the two is in the initial treatment of the grain. Dry milling plants cost less to build and produce higher yields of ethanol (2.7 gallons per bushel of corn), but the value of the co-products is less. The value of

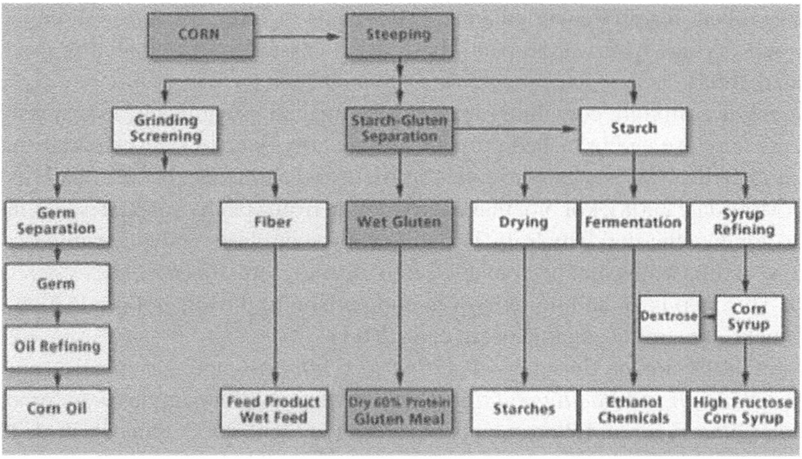

Fig. 5.8 Ethanol Production Process—Wet Milling http://www.ethanolrfa.org/pages/how-ethanol-is-made. Reproduced with permission

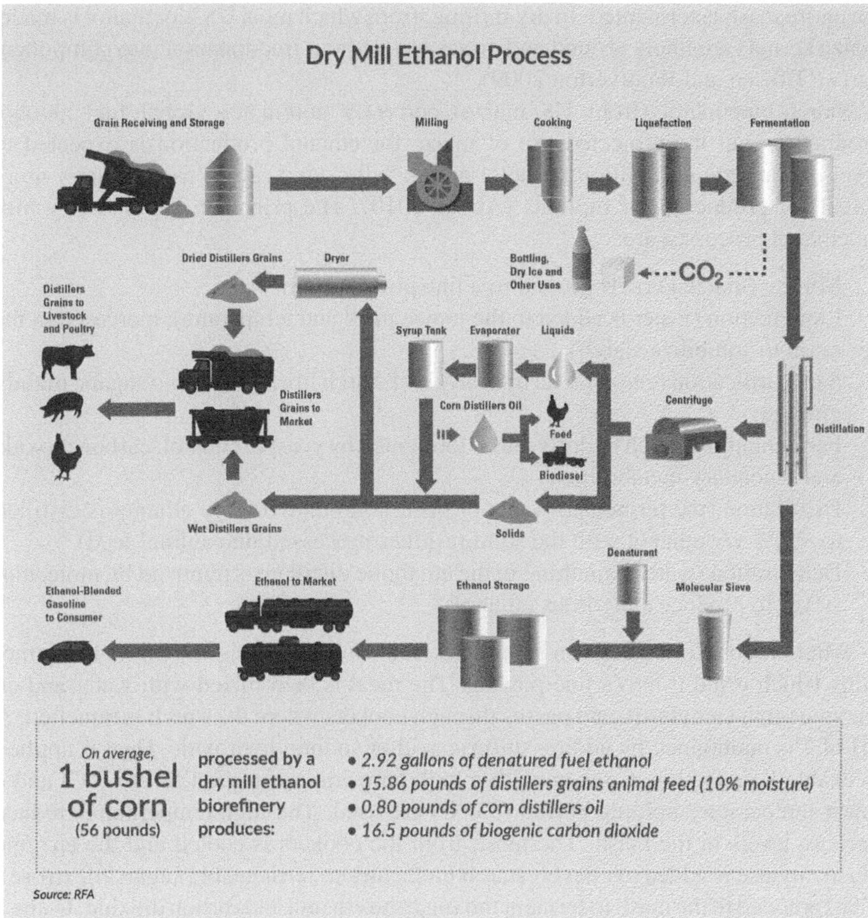

Fig. 5.9 The Ethanol Production Process—Dry Milling RFA (2015). Reproduced with permission https://ethanolrfa.org/wp-content/uploads/2015/09/Ethanol-Industry-Outlook-2015.pdf

corn as a feedstock for ethanol production is due to the large amount of carbohydrates specifically starch present in corn. In wet milling, maize kernels are soaked in water (or dilute acid) to separate the cereal into starch, gluten, protein, oil and fibre prior to starch conversion to ethanol.

The wet-milling operation is more elaborate because the grain must be separated into its components. After milling, the corn is heated in a solution of water and sulfur dioxide for 24–48 h to loosen the germ and the hull fiber. The germ is then removed from the kernel, and corn oil is extracted from the germ. The remaining germ meal is added to the hulls and fiber to form corn gluten feed. A high-protein portion of the kernel called gluten is separated and becomes corn gluten meal which is used for animal feed. In wet milling, only the starch is fermented, unlike dry milling, when the entire mash is fermented. In dry milling, from which most US bioethanol is made, maize kernels are finely ground and processed without fractionation into component parts (O'Brien and Woolverton 2009).

Starch-bioethanol (from US maize) currently dominates global fuel alcohol production, but the projected use of maize for ethanol production is expected to level-off (at around 6 billion bushels) unless "idle" land can be used to grow more cereal for production of biofuels (Abbas 2010). The principal stages in dry mill bioethanol processes are:

1. Milling (maize kernels ground to a fine powder or meal)
2. Liquefaction (water is added to the maize meal and temperature increased in the mash to solubilize starch)
3. Saccharification (enyzymatic hydrolysis of starch liberates simple sugars, mainly glucose)
4. Fermentation (starch hydrolysate is fermented by yeast to ethanol, carbon dioxide and secondary metabolites)
5. Distillation (the fermented wash, or beer, at around 10%v/v ethanol is distilled to ~96% v/v ethanol with the solid residues processed into animal feed)
6. Dehydration (water remaining in the ethanolic distillate is removed by molecular sieves to produce anhydrous ethanol)

After the corn (or other grain or biomass) is cleaned, it passes first through hammer mills which grind it into a fine powder. The meal is then mixed with water and an enzyme (alpha amylase), and passes through cookers where the starch is liquefied. A pH of 7 is maintained by adding sulfuric acid or sodium hydroxide. Heat is applied to enable liquefaction. Cookers with a high temperature stage (120°-150 °C) and a lower temperature holding period (95 °C) are used. The high temperatures reduce bacteria levels in the mash. The mash from the cookers is cooled and the enzyme glucoamylase is added to convert starch molecules to fermentable sugars (dextrose). Yeast is added to the mash to ferment the sugars to ethanol and carbon dioxide. Using a continuous process, the fermenting mash flows through several fermenters until the mash is fully fermented and leaves the tank. In a batch fermentation process, the mash stays in one fermenter for about 48 h. The fermented mash, now called "beer," contains about 10% alcohol, as well as all the non-fermentable solids from the corn and the yeast cells. The mash is then pumped to the continuous flow, multi-column distillation system where the alcohol is removed from the solids and water. The alcohol leaves the top of the final column at about 96% strength, and the residue mash, called stillage, is transferred from the base of the column to the co-product processing area. The stillage is sent through a centrifuge that separates the coarse grain from the solubles. The solubles are then concentrated to about 30% solids by evaporation, resulting in Condensed Distillers Solubles (CDS) or "syrup." The coarse grain and the syrup are then dried together to produce dried distillers grains with solubles (DDGS), a high quality, nutritious livestock feed. The carbon dioxide released during fermentation is captured and sold for use in carbonating soft drinks and beverages and the manufacture of dry ice. Drying the distillers grain accounts for about 1/3 of the plants energy usage. The alcohol then passes through a dehydration system where the remaining water is removed. Most plants use a molecular sieve to capture the last bit of water in the ethanol. The alcohol at this stage is called *anhydrous* (pure, without water) ethanol and is approximately 200 proof. Ethanol that is used for fuel is then denatured with a small amount (2–5%) of some product, like gasoline, to make it unfit for human consumption.

Starch is an alpha-polysaccahride comprising D-glucose monomers arranged in two basic formats: amylose and amylopectin (Fig. 5.10a, b) and plant starches generally contain 10–25% amylose and 75–90% amylopectin (depending on the biomass source). Industrial enzymes used as processing aids in starch-to-ethanol bioconversions are produced by microorganisms (bacteria such as *Bacillus* spp. and fungi such as *Aspergillus* spp.) grown in closed fermentation tanks by major enzyme producers example Novozymes, Genencor. The industrial production and purification of amylolytic enzymes for bioethanol production have been reported (Nair et al. 2008). For conversion of starch to ethanol by *S. cerevisiae* yeast, it has to be depolymerised to its constituents such as glucose and maltose. In traditional beverage fermentation industries such as brewing, this is partly done using endogenous enzymes, mostly alpha and beta amylase enzymes which are present in malted barley. But, for production of bioethanol, more complete starch hydrolysis is needed

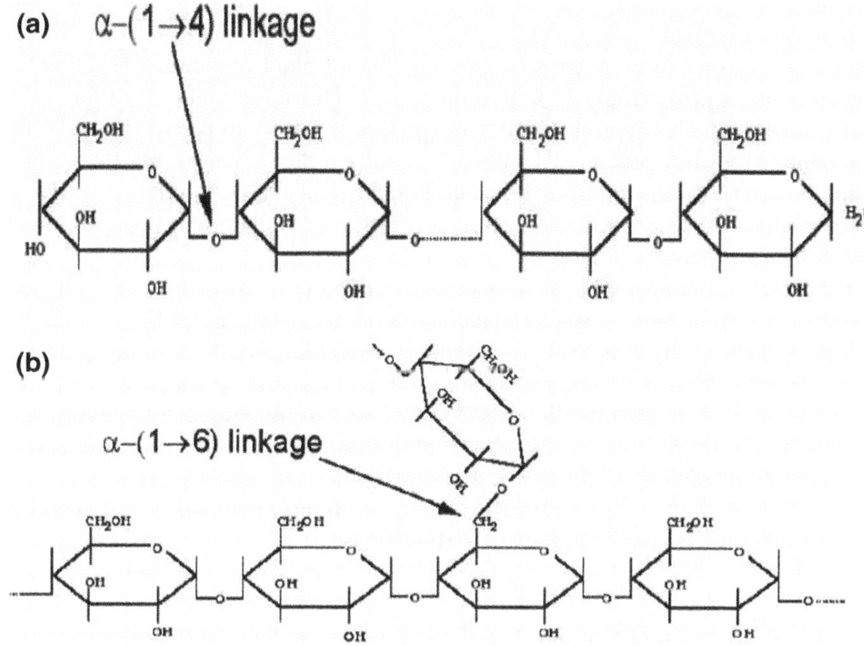

Fig. 5.10 **a** Structure of amylase. Based on http://www.gtconsult.com.br/ingles/artigos/What_is_Starch.pdf, http://www.scientificpsychic.com/fitness/carbohydrates1.html. **b** Structure of Amylopectin. Based on http://www.gtconsult.com.br/ingles/artigos/What_is_Starch.pdf, http://www.scientificpsychic.com/fitness/carbohydrates1.html

and this is done using extracellular amylolytic enzymes derived from microorganisms including debranching enzymes such as amyloglucosidasese.

The production of ethanol is an example of how science, technology, agriculture, and allied industries must work in harmony to change a farm product into a fuel. Ethanol plants receive the large quantities of corn they need by truck, rail, or barge. The corn is cleaned, ground, and blown into large tanks where it is mixed into a slurry of corn meal and water. Enzymes are added and exact acidity levels and temperatures are maintained, causing the starch in the corn to break down–first into complex sugars and then into simple sugars.

New technologies have changed the fermentation process. In the beginning it took several days for the yeast to work in each batch. A new, faster and less costly method of continuous fermentation has been developed. Plant scientists and geneticists are also involved. They have been successful in developing strains of yeast that can convert greater percentages of starch to ethanol. Scientists are also developing enzymes that will convert the complex sugars in biomass materials to ethanol. Cornstalks, wheat and rice straw, forestry wastes and switchgrass all show promise as future sources of ethanol.

In modern ethanol production, for every bushel of corn that is processed, one-third is returned to the livestock feed market. That's because ethanol production requires only the starch portion of a corn kernel. The remaining protein, fat, fiber and other nutrients are returned to the global livestock and poultry feed markets (RFA 2007a). Thus, every bushel of corn processed by an ethanol plant produces 2.8 gallons of ethanol—and approximately 17 lb of animal feed. This high-quality feed for cattle, poultry and pigs isn't a byproduct of ethanol production; it's a co-product. In 2017, the U.S. ethanol industry generated a record 41.4 million metric tons (mmt) of distillers grains, gluten feed and gluten meal. In addition, the industry also produced nearly 3.6 billion pounds of corn distillers oil, used as a feed ingredient or biodiesel feedstock (Fig. 5.11) (RFA 2018).

This level of output will make it necessary to find new markets and uses for co-products. New uses being considered include food, fertilizer and cat litter. While the majority of feed is dried and sold as Distillers Dried Grains with Solubles (DDGS), approximately 20-25% is fed wet locally, reducing energy costs associated with drying as well as transportation costs. Ethanol wet mills produced approximately 430,000 metric tons of corn gluten meal, 2.4 million metric tons of corn gluten feed and germ meal, and 565 million pounds of corn oil.

Figures 5.12 and 5.13 show Distillers grain consumption by species and Distillers grains production by type respectively (RFA 2018).

5.2.2.1 Production of Ethanol From Lignicellulosic Raw Materials

Production of bioethanol from lignocelluloses is attractive and sustainable because these are renewable and uncompetitive with food crops. Moreover, the use of bioethanol produced from lignocellulosic raw materials is related to significant reduction of green house gas emission (Binod et al. 2010). Lignocellulosic raw materials

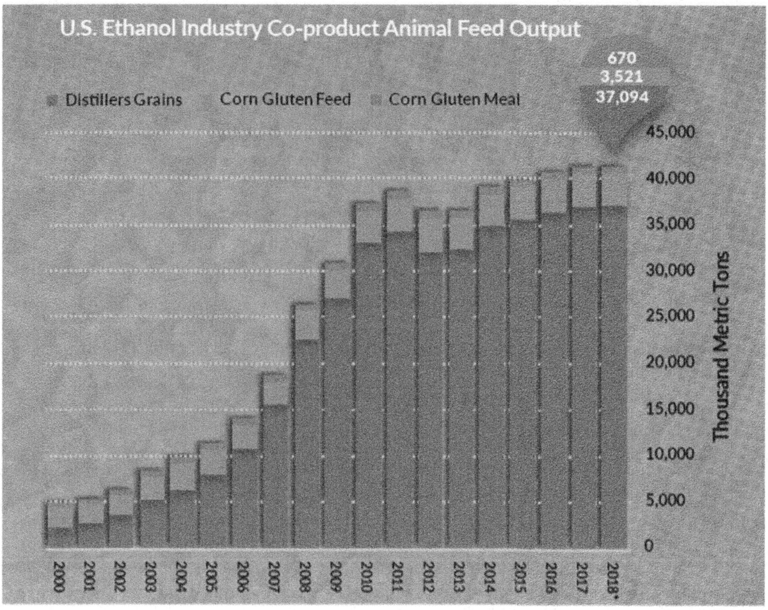

Fig. 5.11 Production of US ethanol feed coproducts. RFA (2018). Reproduced with permission

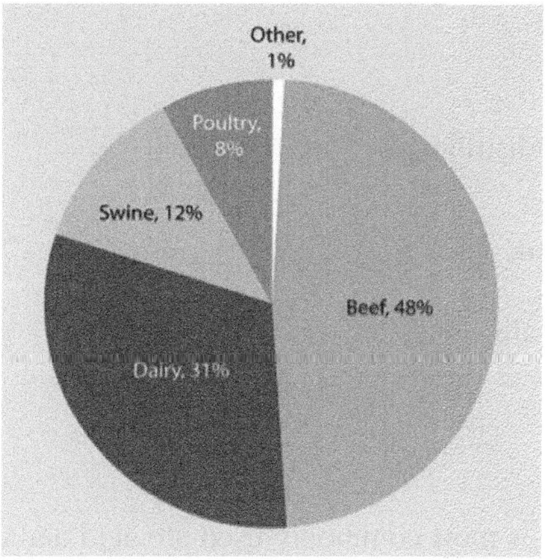

Fig. 5.12 Distillers grain consumption by species. RFA (2018). Reproduced with permission

are almost equally distributed on the Earth, in comparison to the fossil resources. These raw materials provide security of supply by using domestic energy sources (Soccol et al. 2011). These materials can be obtained from different residues or directly harvested from forest and it is less costly in comparison to sugar or starch containing raw materials, requiring full agricultural breeding strategy. Raw materials

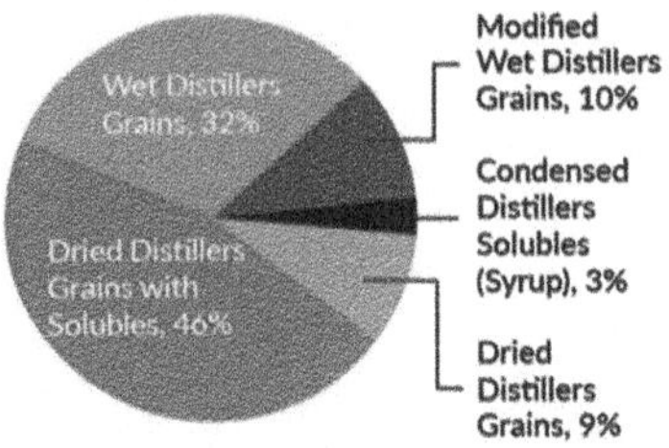

Fig. 5.13 Distillers grains production by type. RFA (2018). Reproduced with permission

containing lignocellulose for bioethanol production form six major groups (Quintero et al. 2011):

crop residues (cane and sweet sorghum bagasse, corn stover, different straw types, rice hulls, olive stones and pulp)

hardwood (aspen, poplar)

softwood (pine, spruce)

cellulose wastes (e.g. waste paper and recycled paper sludge)

herbaceous biomass (alfalfa hay, switchgrass and other types of grasses)

municipal solid wastes

Bioconversion of lignocellulosics to ethanol consists of four main steps:

- Pretreatment
- Hydrolysis
- Fermentation
- Product separation/distillation

5.2.2.2 Pretreatment

Pl. see Chap. 6 for details.

5.2.2.3 Hydrolysis

The hydrolysis methods most commonly used are acid and enzymatic. Both dilute and concentrated acid is used. Dilute acid treatment is employed for the degradation of hemicellulose leaving lignin and cellulose network in the substrate. Other treatments are alkaline hydrolysis or microbial pretreatment with white-rot fungi (*Phaenerochate chrysosporium*, *Cyathus stercoreus*, *Cythus bulleri* and *Pycnoporous cinnabarinus* etc.) preferably act upon lignin leaving cellulose and hemicellulose network in the residual portion. However, during both treatment processes, a considerable amount of carbohydrates are also degraded, hence the carbohydrate recovery is not satisfactory for ethanol production. The dilute acid process is conducted under high temperature and pressure and has reaction time in the range of seconds or minutes. The concentrated acid process uses relatively mild temperatures, but at high concentration of sulfuric acid and a minimum pressure involved, which only creates by pumping the materials from vessel to vessel. Reaction times are typically much longer than for dilute acid. In dilute acid hydrolysis, the hemicellulose fraction is depolymerized at lower temperature than the cellulosic fraction. Dilute sulfuric acid is mixed with biomass to hydrolyse hemicellulose to xylose and other sugars. Dilute acid is interacted with the biomass and the slurry is held at temperature ranging from 120–220 °C for a short period of time. Thus hemicellulosic fraction of plant cell wall is depolymerised and will lead to the enhancement of cellulose digestibility in the residual solids (Nigam 2002; Sun and Cheng 2002; Dien et al. 2006; Saha et al. 2005). Dilute acid hydrolysis has some limitations. If higher temperatures (or longer residence time) are applied, the hemicelluosic derived monosaccharides will degrade and give rise to fermentation inhibitors like furan compounds, weak carboxylic acids and phenolic compounds (Olsson and Hahn-Hägerdal 1996; Klinke et al. 2004; Larsson et al. 1999). These fermentation inhibitors are known to affect the ethanol production performance of fermenting microorganisms (Chandel et al. 2007). In order to remove the inhibitors and increase the hydrolysate fermentability, several chemicals and biological methods have been used. These methods include overliming (Martinez et al. 2000), charcoal adsorption (Chandel et al. 2007), ion exchange (Nilvebrant et al. 2001), detoxification with laccase (Martín et al. 2002; Chandel et al. 2007), and biological detoxification (Lopez et al. 2004). The detoxification of acid hydrolysates has been shown to improve their fermentability; however, the cost is often higher than the benefits achieved (Palmqvist and Hahn-Hagerdal 2000; von Sivers and Zacchi 1995). Dilute acid hydrolysis is carried out in two stages-first-stage and second-stage.

Enzymatic hydrolysis using cellulases does not generate inhibitors and the enzymes are very specific for cellulose. Cellulases, mainly derived from fungi like *Trichoderma reesei* and bacteria like *Cellulomonas fimi*, are a mixture of at least three different enzymes:

(1) endoglucanase which attacks regions of low crystallinity in the cellulose fiber, creating free chain-ends;
(2) exoglucanase or cellobiohydrolase which degrade the molecule further by removing cellobiose units from the free chain-ends
(3) β-glucosidase which hydrolyzes cellobiose to produce glucose, in much smaller amounts. Cellulase enzyme, however, has been projected as a major cost contributor to the lignocellulose-to-ethanol technology. The main challenges are the low glucose yield and high cost of the hydrolysis process. Cellulase costs around 15–20 cents per gallon ethanol as compared to only 2–4 cents per gallon ethanol for amylases used in the starch-to-ethanol process, and ethanol production from sugarcane and molasses bypasses this cost entirely. The cellulase requirement is also much higher than for amylases. The complex three-dimensional (crystalline) structure of cellulose slows down the rate of hydrolysis dramatically; starch hydrolysis by amylases is 100 times faster than cellulose hydrolysis by cellulases under industrial processing conditions.

The US Department of Energy (DOE), signed contracts (worth \$17 million and \$12.3 million, respectively) with Genencor International and Novozymes Inc., to increase cellulase activity and bring down the cost of cellulose enzymes. Novozymes has launched CellicTM, a new class of cellulases, which is claimed to be more cost effective and four times more active compared to the previously produced cellulases (www.bioenergy.novozymes.com). On similar lines, Genencor International has announced the launch of Accellerase 1500. This is claimed to have significantly improved formulation and higher activity resulting in higher ethanol yields and robust operation in a wider variety of processes. The product is based on Genencor's patented gene expression Technology.

Enzymatic hydrolysis is highly specific and can produce high yields of relatively pure glucose syrups, without generation of glucose degradation products. Utility costs are low, as the hydrolysis occurs under mild reaction conditions. The process is compatible with many pre-treatment options, although purely physical methods are typically not adequate (Graf and Koehler 2000; Sun and Cheng 2002). Many experts see enzymatic hydrolysis as key to cost-effective ethanol production in the long run. Although acid processes are technically more mature, enzymatic processes have comparable projected costs and the potential of cost reductions as technology improves (Lynd et al. 1999).

The rate and extent of enzymatic hydrolysis of lignocellulosic biomass depends on enzyme loadings, hydrolysis periods, and structural features resulting from pretreatments. The extent of cellulose conversion is dictated by the degree of attachment of the cellulase enzymes to the three-dimensional cellulose surface. Crystallinity of cellulose affects the initial rate of enzymatic hydrolysis while presence of lignin and acetyl groups limits the extent of cellulose hydrolysis due to unproductive binding of cellulase with lignin and the acetyl groups (Ghose and Bisaria 1979). Thus, an efficient pretreatment strategy that decreases cellulose crystallinity and removes lignin to the maximum extent can significantly decrease hydrolysis time as well as cellulase loading. The problem of non-specific binding of cellulases to lignin can be circumvented by adding non-ionic surfactants like Tween 20. It has been shown that surfactant addition can increase the ethanol yield by 8% and reduce cellulase loading by as much as 50%, while maintaining a constant ethanol yield. The proposed mechanism for this enhanced enzyme performance is the adsorption of the surfactant onto lignin surface, thereby preventing unproductive binding of the enzymes to lignin. The adsorption also helps in retaining the enzymes for recycle. Cheaper alternatives to non-ionic surfactants like Tween 20 can reduce the process cost considerably.

Another promising approach to enhancing enzymatic hydrolysis of pre-treated biomass is the use of ultrasonic energy. The cavitational effect of ultrasound leads to biomass swelling and fragmentation, increasing the accessibility of cellulases to cellulosic substrate (Ebringerova and Hromadkova 2002). Intermittent ultrasonication of biomass before and after enzymatic hydrolysis has been shown to significantly increase the rate of reaction. Approximately 20% increase in ethanol yield and 50% decrease in enzyme loading were observed on intermittent exposure of simultaneous saccharification and fermentation processes of mixed waste office paper to ultrasonic energy under selected conditions (Banerjee et al. 2010). *Saccharomyces cerevisiae* is by far the most commonly used microbial species for industrial ethanol production from sugar- and starch-based raw materials, and is well adapted to the industrial scenario. It produces ethanol with stoichiometric yields and tolerates a wide spectrum of inhibitors and elevated osmotic pressure. *Zymomonas mobilis* has been projected as the future ethanologen due to its high ethanol tolerance (up to 14% v/v), energy efficiency, high ethanol yield (up to 97% of theoretical), and high ethanol productivity. Complete substrate utilization is a pre-requisite for economically competitive lignocellulosic fermentation processes. The native ethanologen *S. cerevisiae* is capable of fermenting only hexoses, and cannot utilize pentoses like xylose, which is the main component of the hemicellulosic fraction of lignocellulose, and can contribute to as much as 30% of the total biomass. *Z. mobilis* can only utilize glucose, fructose, and sucrose. Expanding the substrate range of whole-cell biocatalysts will greatly contribute to the economic feasibility of bioethanol production from renewable feedstock. The essential traits of a good lignocellulose-to-ethanol bioconverter are:

- Utilization of both hexoses and pentoses
- High ethanol yields and productivity
- Minimum byproduct formation
- High ethanol tolerance
- Tolerance to other inhibitors formed during biomass pre-treatment and hydrolysis.

Four industrial benchmarks for ethanologenic strain development, which have the greatest influence on the price of lignocellulosic ethanol are (Banerjee et al. 2010):

- Process water economy
- Inhibitor tolerance
- Ethanol yield
- Specific ethanol productivity.

The inability of *S. cerevisiae* and *Z. mobilis* to utilize pentose sugars necessitates the search for pentose utilizing strains. Two groups of micro-organisms, i.e., enteric bacteria and some yeasts, are able to ferment pentoses, but with low ethanol yields (Bajpai and Margaritis 1982). Furthermore, xylose fermenting yeasts (*Pachysolen tannophilu s*, *Candida shehatae*, and *Pichia stipitis*) are sensitive to high concentrations of ethanol ($\geq$40 g/l), require micro-aerophilic conditions, are highly sensitive to inhibitors, and are not capable of fermenting xylose at low pH.

Table 5.2 shows some popular microorganisms for bioethanol production.

Due to the lack of a natural microorganism for efficient fermentation of lignocellulose-derived substrates, there has been emphasis on constructing an efficient organism through metabolic engineering of different organisms (Banerjee et al. 2010). Metabolic engineering, by virtue of the recent molecular biology tools, has generated recombinant organisms displaying attractive features for the bioconversion of lignocelluloses to ethanol. The three most promising microbial species that have been developed by metabolic engineering in the last two decades are *S. cerevisiae*, *Z. mobilis*, and *Escherichia coli*. The recombinant organisms developed for

Table 5.2 Microorganisms used for bioethanol production

	Saccharomyces cerevisiae	Pichia stipitis	Escherchia coli	Zymomons mobilis
D-Glucose fermentation	+	+	+	+
Other hexose utilization (D-galactose and D mannose)	+	+	+	–
Pentose utilization (D-xylose and L arabinose)	–	+	+	–
Direct hemicellulose utilization	–	w	–	–
Anaerobic fermentation	+	w	–	+
Mixed-product	w	w	+	w
High ethanol productivity (from glucose)	+	w	–	+
Ethanol tolerance	+	w	w	w
Tolerance to lignocellulose-derived inhibitors	+	w	w	w
Acidic pH range	+	w	–	–

+, positive; –, negative; w, weak
Based on Tran et al. (2019), Girio et al. (2010)

lignocellulose-to-ethanol process have been extensively reviewed in earlier works. Significant work on the metabolic engineering of *E. coli* has been completed. The incorporation and expression of pyruvate decarboxylase and alcohol dehydrogenase II genes from *Z. mobilis* into *E. coli* under the control of a common (*lac*) promoter results in high ethanol yield due to co-fermentation of hexose and pentose sugars. Xylose isomerase (XI), encoded by the *xylA* gene, catalyzes the isomerization of xylose to xylulose in bacteria and some fungi. Recent developments to improve XI activity and heterologous expression in *S. cerevisiae* through adaptation and metabolic manipulation have proven to be successful with ethanol yields approaching the theoretical maximum. Genes *xylA* and *xylB*, part of the xyl operon of *E. coli*, were introduced into *Pseudomonas putida* S12 under the transcriptional control of the constitutive *tac* promoter. This recombinant strain, after further laboratory evolution, could efficiently utilize the three, most-abundant sugars in lignocellulose, glucose, xylose, and arabinose, as sole carbon sources. Whereas recombinant laboratory strains are useful for evaluating metabolic engineering strategies, they do not possess the robustness required in the industrial context. The strategies of producing industrial pentose fermenting strains involve the introduction of the initial xylose and arabinose utilization pathways and adaptation strategies, including random mutagenesis, evolutionary engineering, and breeding.

5.2.2.4 Fermentation

Fermentation of the hydrolyzed biomass can be carried out in a variety of different process configurations:

- Separate hydrolysis and fermentation (SHF)
- Simultaneous saccharification and fermentation (SSF)
- Simultaneous saccharification and co-fermentation (SSCF)

Enzymatic hydrolysis performed separately from fermentation step is known as separate hydrolysis and fermentation (SHF) (Wingren et al. 2003). The separation of hydrolysis and fermentation offers various processing advantages and opportunities. It enables enzymes to operate at higher temperature for increased performance and fermentation organisms to operate at moderate temperatures, optimizing the utilization of sugars. The most important process improvement made for the enzymatic hydrolysis of biomass is the introduction of simultaneous saccharification and fermentation (SSF), which has been improved to include the co-fermentation of multiple sugar substrates (Wingren et al. 2003). This approach combines the cellulase enzymes and fermenting microbes in one vessel. This enables a one-step process of sugar production and fermentation into ethanol. Simultaneous saccharification of both carbon polymer, cellulose to glucose; and hemicellulose to xylose and arabinose; and fermentation is carried out by recombinant yeast or the organism which has the ability to utilize both C5 and C6 sugars. According to Alkasrawi et al. (2006) the mode of preparation of yeast must be carefully considered in SSF designing. A more robust strain will give substantial process advantages in terms of higher solid loading and possibility to recirculate the process stream, which results in increased energy demand and reduced fresh water utilization demand in process. Adaptation of yeast to the inhibitors present in the medium is an important factor for consideration in the design of SSF process. SSF combines enzymatic hydrolysis with ethanol fermentation to keep the concentration of glucose low. The accumulation of ethanol in the fermenter does not inhibit cellulase action as much as high concentration of glucose; so, SSF is good strategy for increasing the overall rate of cellulose to ethanol conversion (Tanaka 2006; Krishna et al. 2001; Kroumov et al. 2006). SSF gives higher ethanol yield while requiring lower amounts of enzyme because end-product inhibition from cellobiose and glucose formed during enzymatic hydrolysis is relieved by the yeast fermentation (Banat et al. 1998). However, it is not feasible for SSF to meet all the challenges at industrial level due to its low rate of cellulose hydrolysis and most microorganisms employed for ethanol fermentation can not utilize all sugars derived after hydrolysis. To overcome this problem, the cellulolytic enzyme cocktail should be more stable in wide range of pH and temperature. Also, the fermenting microorganisms should be able to ferment a wide range of C5 and C6 sugars. Matthew et al. (2005) have found some promising ethanol producing bacteria viz. recombinant *E. coli* KO11, *Klebsiella oxytoca*, and *Zymomonas mobilis* for industrial exploitation. SSF process has now improved after including the co-fermentation of multiple sugar substrates present in the hydrolysate. This new variant of SSF is known as simultaneous saccharification and co- fermentation (SSCF) (Wyman et al. 2005). SSF and

SSCF are preferred over SHF, since both operations can be performed in the same tank resulting in lower cost, higher ethanol yield and shorter processing time. The most upgraded form of biomass to ethanol conversion is consolidated bioprocessing (CBP)—featuring cellulose production, cellulose hydrolysis and fermentation in one step – is a highly integrated approach with outstanding potential (Lynd et al. 2005). It has the potential to provide the lowest cost route for biological conversion of cellulolosic biomass to ethanol with high rate and desired yields.

Direct microbial conversion is a method of converting cellulosic biomass to ethanol in which both ethanol and all required enzymes are produced by a single microorganism. The potential advantage of direct microbial conversion is that a dedicated process step for the production of cellulase enzyme is not necessary. Cellulase enzyme production (or procurement) contributes significantly to the cost involved in enzymatic hydrolysis process. However, direct microbial conversion is not considered the leading process alternative. This is because there is no robust organism available that can produce cellulases or other cell wall degrading enzymes in conjunction with ethanol with a high yield. Singh and Kumar (1991) found that several strains of *Fusarium oxysporum* have the potential for converting not only D-xylose, but also cellulose to ethanol in a one-step process. Distinguishing features of *F. oxysporum* for ethanol production in comparison to other organisms are identified. These include the advantage of in situ cellulase production and cellulose fermentation, pentose fermentation, and the tolerance of sugars and ethanol. The main disadvantage of *F. oxysporum* is its slow conversion rate of sugars to ethanol as compared to yeast.

The comparison of the process conditions and performance of the different cellulose hydrolysis process shows that the dilute acid process has a low sugar yield (50—70% of the theoretical maximum). The enzymatic hydrolysis has currently high yields (75—85%) and improvements are still projected (85—95%). Moreover, refraining from using acid may be better for the economics (cheaper construction materials, cutting operational costs), and the environment (no gypsum disposal).

The sugar syrup obtained after cellulosic hydrolysis is used for ethanol fermentation. A variety of micro-organism—bacteria, yeast, or fungi–ferment sugars to ethanol under oxygen-free conditions. They do so to obtain energy and to grow. According to the reactions, the theoretical maximum yield is 0.51 kg ethanol and 0.49 kgcarbon dioxide per kg sugar.

Methods for C6 sugar fermentation were already known. The ability to ferment pentoses along with hexoses is not widespread among microorganisms. *S. cereviseae* is capable of converting only hexose sugars to ethanol. The most promising yeasts that have the ability to use both C5 and C6 sugars are *Pichia stipitis*, *Candida shehatae* and *Pachysolan tannophilus*. However, ethanol production from sugars derived from starch and sucrose has been commercially dominated by the yeast *S. cereviseae* (Tanaka 2006). Thermotolerant yeast could be more suitable for ethanol production at industrial level. In high temperature process, energy savings can be achieved through a reduction in cooling costs. Considering this approach, Sree et al. (2000) developed solid state fermentation system for ethanol production from sweet sorghum and potato employing a thermotolerant *S. cereviseae* strain.

Currently the research is focusing on developing recombinant yeast, which can greatly improve the ethanol production yield by metabolizing all form of sugars, and reduce the cost of operation. The efforts have been made by using two approaches. The first approach has been to genetically modify the yeast and other natural ethanologens additional pentose metabolic pathways. The second approach is to improve ethanol yields by genetic engineering in microorganisms that have the ability to ferment both hexoses and pentoses (Jeffries and Jin 2000; Dien et al. 2003). Jeffries and Jin (2000) compiled the developments towards the genetic engineering of yeast metabolism and concluded that strain selection through mutagenesis, adaptive evolution using quantitative metabolism models may help to further improve their ethanol production rates with increased productivities. Piskur et al. (2006) showed the developments in comparative genomics and bioinformatics to elucidate the high ethanol production mechanism from *Saccharomyces* sp. Though new technologies have greatly improved bioethanol production yet there are still a lot of problems that have to be solved. The major problems include: maintaining a stable performance of genetically engineered yeast in commercial scale fermentation operation (Ho et al. 1998, 1999), developing more efficient pre-treatment technologies for lignocellulosic biomass, and integrating optimal component into economic ethanol production system (Dien et al. 2000). Fermentation can be performed as a batch, fed batch or continuous process. The choice of most suitable process will depend upon the kinetic properties of microorganisms and type of lignocellulosic hydrolysate in addition to process economics aspects. Traditionally, ethanol has been produced batch wise. At present, nearly, all of the fermentation ethanol industry uses the batch mode. In batch fermentation, the microorganism works in high substrate concentration initially and a high product concentration finally (Olsson and Hahn-Hägerdal 1996). The batch process is a multi-vessel process, allows flexible operation and easy control over the process. Generally batch fermentation is characterized by low productivity with an intensive labour. For batch fermentation, elaborate preparatory procedures are needed; and because of the discontinuous start up and shut down operations, high labour costs are incurred. This inherent disadvantage and the low productivity offered by the batch process have led many commercial operators to consider the other fermentation methods like fed batch fermentation, continuous fermentation, immobilized cells etc. In fed batch fermentation the microorganism works at low substrate concentration with an increasing ethanol concentration during the course of fermentation process. Fed batch cultures often provide better yield and productivities than batch cultures for the production of microbial metabolites. For practical reasons, therefore, some continuous operations have been replaced by fed batch process. Keeping the low feed rate of substrate solution containing high concentration of fermentation inhibitors such as furfural, hydroxymethyl furfural and phenolics, the inhibitory effect of these compounds to yeast can be reduced. Complete fermentation of an acid hydrolysate of spruce, which was strongly inhibiting in batch fermentation, has been achieved without any detoxification treatment (Taherzadeh 1999). The productivity in fed batch fermentation is limited by the feed rate which, in turn, is limited by the cell mass concentration. The specific ethanol productivity has also been reported to decrease with increasing cell mass concentration (Palmqvist et al.

1996). Ideally, the cell density should be kept at a level providing maximum ethanol productivity and yield. Continuous fermentation can be performed in different kind of bioreactors—stirred tank reactors (single or series) or plug flow reactors. Continuous fermentation often gives a higher productivity than batch fermentation, but at low dilution rates which offers the highest productivities. Continuous operation offers ease of control and is less labor intensive than batch operation. However, contamination is more serious in this operation. Since the process must be interrupted, all the equipments must be cleaned, and the operation started again with the growth of new inoculum. The continuous process eliminates much of the unproductive time associated with cleaning, recharging, adjustment of media and sterilization. A high cell density of microbes in the continuous fermenter is locked in the exponential phase, which allows high productivity and overall short processing of 4–6 h as compared to the conventional batch fermentation (24–60 h). This results in substantial savings in labour and minimizes investment costs by achieving a given production level with a much smaller plant. A limitation to continuous fermentation is the difficulty of maintaining high cell concentration in the fermenter. The use of immobilized cells circumvents this difficulty. Immobilization by adhesion to a surface (electrostatic or covalent), entrapment in polymeric matrices or retention by membranes has been successful for ethanol production from hexoses (Godia et al. 1987).

The applications of immobilized cells have made a significant progress in fuel ethanol production technology (Najafpour 1990; Abbi et al. 1996; Sree et al. 2000; Yamada et al. 2002). Immobilized cells offer rapid fermentation rates with high productivity—that is, large fermenter volumes of mash put through per day, without risk of cell washout. In continuous fermentation, the direct immobilization of intact cells helps to retain cells during transfer of broth into collecting vessel. Moreover, the loss of intracellular enzyme activity can be kept to a minimum level by avoiding the removal of cells from downstream products. Immobilization of microbial cells for fermentation has been developed to eliminate inhibition caused by high concentration of substrate and product and also to enhance ethanol productivity and yield.

5.2.2.5 Recovery of Ethanol

The product stream from fermentation, is a mixture of ethanol, cell mass and water. In this flow, ethanol from cellulosic biomass has likely lower product concentrations ($\leq$ 5 wt%) than in ethanol from corn. The maximum concentration of ethanol tolerated by the micro-organisms is about 10 wt% at 30 °C but decreases with increasing temperature. To maximize cellulase activity, the operation is rather at maximum temperature (37 °C), since the cost impact of cellulase production is high relative to distillation (Lynd 1996). On the processing side, slurries become difficult to handle when containing over 15 wt% solids, which also corresponds to 5% ethanol (two thirds carbohydrates, and < 50 wt% conversion) (Lynd 1996). The first step is to recover the ethanol in a distillation or beer column, where most of the water remains with the solids part. The product (37% ethanol) is then concentrated in a rectifying column to a concentration just below the azeotrope (95%) (Wooley et al. 1999). Hydrated ethanol can be employed in E95 ICEVs (Wyman et al. 1993) or in FCVs (requires onboard reforming), but for mixtures with gasoline water-free (anhydrous) ethanol is required. One can further distill in the presence of an entrainer (e.g. benzene), dry desiccants (e.g. corn grits), or use pervaporation or membranes (Lynd 1996). By recycling between distillation and dehydration, eventually 99.9% of the ethanol in the beer is retained in the dry product (Wooley et al. 1999).

The main solid residual from the process is lignin. Its amount and quality differs with feedstock and the applied process. Production of co-products from lignin, such as high-octane hydrocarbon fuel additives, may be important to the competitiveness of the process (US DOE 2003). Lignin can replace phenol in the widely used phenol formaldehyde resins. Both production costs and market value of these products are complex. In corn based ethanol plants, the stillage (20% protein) is very valuable as animal feed.

5.2.2.6 Recirculation of Process Stream

The water consumption is reduced by recirculating process streams for use in the washing and hydrolysis steps (Palmqvist and Hahn-Hagerdal 2000). Recirculating part of the dilute ethanol stream from the fermenter can increase the ethanol concentration in the feed to the distillation stage. However, computer simulations have shown that recirculation of streams leads to the accumulation of nonvolatile inhibitory compounds (Galbe and Zacchi 1992; Palmqvist et al. 1996). To increase the ethanol productivity, cell recycling has been employed by several workers while retaining the simplicity of the batch process. Cell recycling generally does not increase the sugar consumption or ethanol production but the time required for the fermentation can be reduced by 60–70%. Schneider (1989) observed a reduction in ethanol production after third cell cycle and suggested the decrease in ethanol production was due to the limitations of oxygen and sugar as a result of an increase in cell density.

5.2.3 Use of Pulp and Paper Industry Wastes for Ethanol Production

5.2.3.1 Kraft Pulp

Several studies have been undertaken on production of ethanol from Kraft pulp (Branco et al. 2019).

Monrroy et al. (2012) studied the SSF of Kraft pulps from Eucalyptus globules. These pulps were subjected to pretreatment under diverse conditions. The maximum ethanol yield was 0.202 g/g dwt and concentration of ethanol ranged between 30–38 g/l.

Ko et al. (2012) examined SSF of unbleached Kraft pulps of Eucalyptus globules for ethanol production with a different strain of *S. cerevisiae*. The concentration of ethanol and yield were lower. When unbleached Acacia pulp produced by Kraft Process was used, the maximum ethanol yield was 0.045 g/g dwt and ethanol concentration was 5.88 g/l. Better results were obtained when pretreatment was done using steam explosion for both type of woods.

Buzała et al. (2017) studied ethanol production from Kraft pulps from different sources using SHF. Ethanol yield was 0.11 to 0.14 g/g dwt with five hardwood unbleached pulps. For softwood pine unbleached pulp ethanol yield was 0.02 g g/g dwt and for bleached pine pulp, ethanol yield was 0.20 g/g dwt. The lower ethanol yield from unbleached pine pulp was due to the presence of higher amount of extractives in the pulp.

Wistara et al. (2016) studied the SSF of Kraft pulps of Jabon wood having varying lignin content and freeness. Ethanol yield was found in the range of 0.022 and 0.129 g/g dwt. Pulps containing low lignin content and high pulp freeness showed higher ethanol yield.

Edgardo et al. (2008) isolated a thermotolerant strain of *S. cerevisiae* and used it for SSF of bleached Kraft *Pinus radiate* pulp. A ethanol yield of 62% and ethanol concentration of 28 g/L was obtained. With organosolv pretreatment ethanol concentration was low (22 g/L).

Bauer and Gibbons (2012) studied the SSF of Kraft pulp. They used varying enzyme dose for hydrolysis. With *S. cerevisiae*, the maximum values of ethanol yield, ethanol concentration and ethanol productivity were 85.90 ± 5.3%, 17.90 ± 0.99 g/l and 0.25 ± 0.015 g/l/h respectively. With *Candida molischiana,* comparble results were obtained.

Amoah et al. (2017) examined bioethanol production from hardwood unbleached Kraft pulp through consolidated bioprocess. Ethanol yield, ethanol concentration and ethanol productivity were found to be lower.

5.2.3.2 Sulfite Pulping Liquor (SSL)

SSL is a byprduct from sulfite pulping process (Branco et al. 2019). The major constituents of SSL are LS and carbohydrates which result from hemicelluloses. These are pentose and hexose sugars - arabinose, xylose, mannose, galactose, and glucose. These can be fermented. SSL also has degradation products of sugar, such as furfural and hydroxymethyl furfural and uronic acids, methyl glyoxal, methyl alcohol, acetic acid, formaldehyde and extractives (Sjöström 1993). Kraft black liquor is not suitable for ethanol production as sugars are not present. In this mainly lignin and carbohydrate degradation products (for instance hydroxycarboxylic acids, acetic acid, and formic acid), and also some extractives such as turpentine and tall oil and other diverse products are present (Pereira et al. 2013; Ek et al. 2009; Sjöström 1993).

As there is a difference in softwood hemicellulose and hardwood hemicellulose, the sugar composition of SSL is dependent upon the type of wood used for pulping.

SSLs produced from pulping of softwood consist of mostly hexose sugars, whereas SSLs produced from pulping of hardwood present high amount of pentose sugars (Pereira et al. 2013). Moreover SSL is a low price and abundantly available feedstock. In 2015, SSL production was 90 billion liters (Evtuguin 2016; Portugal-Nunes et al. 2015). The biological oxygen demand in SSL is high. This presents a disposal problem which can be solved by utilizing it for the production of value added products such as ethanol, xylitol, polyhydroxyalkanoates, organic acids and fumaric acid (Nigam 2001; Rueda et al. 2015, 2016; Queirós et al. 2014, 2016, 2017; Figueira et al. 2017).

Production of bioethanol from SSLs has been extensively studied, with many research papers published in this area in 1980s and 1990s (Linden and Hahn-Higerdal 1989; Safi et al.1986; Mandenius 1988; Björling and Lindman 1989; Yu et al. 1987; Olsson and Hahn-Hägerdal 1993; Lawford and Rousseau 1993; Marko-Varga et al. 1994; Mohandas et al. 1995; Schneider 1996; Palmqvist et al. 1998).

Many researchers have studied ethanol production from SSL. Portugal-Nunes et al. (2015) investigated the effect of cell immobilization and pH on fermentation of biodetoxified SSL of E. globulus (hardwood) with *S. stipitis*. The concurrent use of cell immobilization and control of pH at 5.5 favored the fermentative metabolism. Ethanol yield, increased by 1.3 times and the volumetric productivity was maintained.

Harner et al. (2015) utlized genome shuffling (a strain improvement technology) for producing mutants of *Pachysolen tannophilus* showing higher tolerance to inhibitors present in HSSL. Among the various genome shuffled strains, GHW301 produced the maximum amount of ethanol, in the diverse media containing HSSL (70 to 100% v/v). Ethanol productivity was 7.4–8.5 g/l/h.

Pereira et al. (2015), used an evolutionary engineering approach. These researchers, obtained a population of *S. stipitis* showing higher tolerance to E. globulus SSL inhibitors. Many isolates adapted to HSSL inhibitors were produced and the best isolate was utilized for the fermentation of HSSL not detoxified, resulting in better performance in comparison to the parent strain. This isolate provided superior results using a two stage aerated fermentation strategy (Henriques et al. 2018).

Takahashi et al. (2015) used anion exchange resins for removing acetic acid from HSSL. Calcium oxide treatment followed by neutralization with carbon dioxide and a two stage strong base ion exchange column removed ~90% of the acetic acid. When the detoxified HSSL was feremented better results were obtained as compared to untreated HSSL. Ethanol productivity was 0.441 g/l/h and yield was 10.6 g/l.

SSL produced from E. Globules when fermented with *S. stipitis*, produced a maximum ethanol concentration of 2.4 g/l. The ethanol yield was 0.24 g/g (Pereira et al. 2012).

Pinel et al. (2011) explored genome shuffling approach for obtaining *S. cerevisiae* mutants showing a higher tolerance to inhibitors in HSSL. The genome shuffled strain R57 was found to be highly productive. Ethanol yield of 80% of the theoretical was achieved.

Bajwa et al. (2010) also used genome shuffling strategy for improving tolerance to HSSL inhibitors. Two mutant strains of *S. stipitis* were examined for ethanol fermentation of HSSL which was not diluted. Ethanol concentration reached ~1.8 g/l.

Deacidification of SSL of E. globulus using *S. cerevisiae* and the subsequent fermentation of the HSSL was examined by Xavier et al. (2010). Ethanol production was not effective most likely because of the presence of polyphenolic inhibitors. These researchers then used the ion-exchange resins for detoxifying SSL of E. globulus. The fermentation of detoxified HSSL produced 8.1 g/l of ethanol. The ethanol yield was 0.49 g/g.

The fermentation of HSSL using *S. stipitis* was studied by Nigam (2001) and Bajwa et al. (2009). Ethanol concentration was found to increase from 6.7 to 20.2 g/l using adapted *P. stipitis* and pretreated HSSL (Nigam 2001). The maximum ethanol yield and productivity were 0.41 g/l and 0.44 g/l/h respectively.

5.2.3.3 Pulp and Paper Industry Sludge (PPMS)

PPMS is generated during pulp and paper production. In the sludge, short fibers are present which are easily hydrolyzed by enzymes into fermentable sugars, and the amount of lignin is less (Branco et al. 2019). "As a result, PPMS requires no pretreatment for the production of ethanol and inhibitors resulting from lignin can be neglected (The major challenge of using PPMS for bioethanol production is the high content in ash, namely $CaCO_3$, which adsorbs to the enzymes and increases the pH of the sludge, hindering enzymatic hydrolysis. In order to surpass this limitation, acidic treatments are frequently used to neutralize $CaCO_3$" (Branco et al. 2019; Guan et al. 2016; Jain et al. 2016; Ridout et al. 2016; Wang et al. 2010; Gurram et al. 2015; Mendes et al. 2014; Marques et al. 2008).

Furthermore, presence of high ash content limits solid loading. This leads to larger reaction vessels which results in higher cost. The water holding capacity and viscosity of PPMS is high. This results in not enough mixing and mass transfer. Fed batch fermentation approach was studied as a approach for beating these challenges (Boshoff et al. 2016; Mendes et al. 2017). Many studies have been undertaken on production of bioethanol from pulp and paper mill sludge.

Schroeder et al. (2017) examined the bioethanol production from sludge obtained from recycled paper sludge. With SHF configuration, it was found that the addition of the hydrolyzed sludge with nutrients resulted in higher production of ethanol. SSF without supplementation resulted in higher ethanol yield in a short time in comparison to SHF. Furthermore, hydrolysis was found to be the rate-limiting step in both types of configurations, most probably because of the interference of the high ash and solid content, as the sludge was not neutrilized.

Mendes et al. (2017) examined bioethanol production from primary sludge using SSF. In batch fermentation mode using a carbohydrate content of 50 g/l, *S. cerevisiae* showed better results than *Kluyveromyces marxianus* which was thermotolerant. Fed batch fermentation resulted in higher solid loading and reduced enzyme load, providing a higher ethanol yield Mendes et al. (2017).

Boshoff et al. (2016) examined SSF of two different PPMS by using fed batch mode. Sludge from production of virgin pulp has high viscosity. Ethanol concentration and yield were lower in comparison to sludge from corrugated recycled paper.

Digestibility, water holding capacity, and viscosity were the main factors affecting this high solid loading SSF in fed batch mode.

Mendes et al. (2016) studied SSF of two primary sludges using batch and fed batch mode. Batch fermentation of both type of sludges produced comparable ethanol concentrations. For the similar enzyme dose, results with fed batch fermentation were lower in comparison to those obtained in batch mode. The sludge containing higher ash content showed better fermentation.

Gurram et al. (2015) examined chemical (with hydrochloric acid), and mechanical de-ashing of primary sludge. It was observed that the enzymatic hydrolysis improved in case of deashed sludges. These researchers also examined batch SHF of diverse types of primary sludges, either chemically deashed or without deashing, when two different accelerants or hydrogen peroxide pretreatment were used with enzymatic hydrolysis. For the four types of sludge, the addition of cationic polyacrylamide accelerant XP10020 during hydrolysis with enzyme showed high ethanol productivity and yield.

Mendes et al. (2014) examined different inorganic and organic acids for neutralizing calcium carbonate in primary sludge. It was found that hydrochloric acid or spent acid were effective, but emitted carbon dioxide. The sludges were then hydrolyzed with enzyme. The hydrolysate was later fermented. These researchers made a comparison of *S. cerevisiae* which ferments hexose and *S. stipitis* which ferments pentose for ethanol production, as sludges which are obtained from mills processing hardwoods consists of a mixture of glucose as well as xylose. *S. stipitis* was found to perform better.

Peng and Chen (2011) explored SHF of primary sludge without subjecting to any pretreatment in a batch process. Optimized approach based on statistical experimental design experiments was used for improving enzymatic hydrolysis performance and achieving successful subsequent fermentation with *S. cerevisiae.*

5.2.4 Converting Pulp and Paper Mills into Biorefineries

"The forest-based sector that utilizes and produces LCB feedstocks is expected to play a major role in the future bioeconomy. The conversion of pulp and paper mills into biorefineries by integrating the ethanol production could valorize the wastes resulting from the mills and diversifying the products, as well as increasing the profitability of pulp and paper industry. One example of this conversion is the reorientation of closed Kraft pulp mills to bioethanol production" (Branco et al. 2019; Pelli et al. 2017; Monrroy et al. 2012; Phillips et al. 2013).

Wu et al. (2014) examined the technoeconomics of repurposing a Kraft mill for ethanol production from loblolly pine.

Fornell et al. (2012) studied the energy efficiency and conducted an economic analysis of a conceptual Kraft pulp mill which was converted into a ethanol plant (Fornell and Berntsson 2012; Fornell et al. 2012).

"The conversion of pulp and paper mills into integrated biorefineries is encouraged by the successful example of Borregaard company, which runs one of the most advanced biorefineries that is currently in operation in Norway. Borregaard begun producing cellulose pulp in 1889, and in 1938, it started to produce ethanol from the hemicelluloses of spruce (softwood), mainly mannose, which were present in the SSL resulting from the pulping process. Today, Borregaard has the capacity to produce 20,000 m^3 of ethanol annually, making it the largest producer of the second generation ethanol. This biorefinery presents the strategy of producing the maximum number of products out of spruce wood, and today products from cellulose, hemicellulose, and lignin are produced in a larger scale. Borregard biorefinery produces cellulose, ethanol, liquid lignin, lignin powder, vanisperse, vanillin, sodium hypochlorite, hydrochloric acid, chlorine, and steam" (Rødsrud et al. 2012; Branco et al. 2019).

Modahl et al. (2015) examined the effect on the environment associated with chemical products of this biorefinery and found that production of ethanol, presents a good performance concerning global warming potential.

"Second generation biorefineries are operating in all regions of the world, bringing far more favorable energy balances to biofuels production than have been previously realized. Substantial displacement of a significant portion of fossil-based liquid fuels has been demonstrated to be a realistic possibility. However, in the face of low petroleum prices, continuing policy support and investment in research and development will be needed to allow biofuels to reach their full potential" (www.dovetailinc.org).

Figure 5.14 shows Global biorefineries (Nguyen et al. 2017).

"As of early fall 2015, 67 biorefineries worldwide were producing second generation ethanol, biodiesel, or aviation biofuel. Over a third of these were operating on

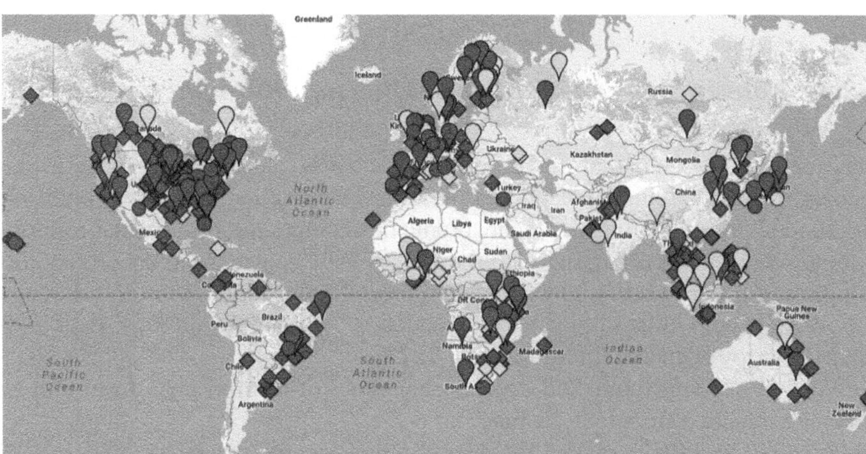

Fig. 5.14 Global biorefineries. Nguyen et al. (2017) Reproduced with permission

Fig. 5.15 Different cultivation systems for algal cultivation. **a** Open ponds. **b** Tubular PBRs. **c** Flat PBRs. **d** Biofilm based PBRs. **e** Fermentor cultivation. **f** Algal biomass cultivation using waste water. Bibi et al. (2017). Reproduced with permission

commercial scale" (www.dovetailinc.org) (Table 5.3). North America has the most commercial second generation plants.

5.3 Algae for Bioethanol Production

Algae are receiving substantial consideration for biofuel production. Algae also address sustainability issues which are mainly associated with food production from crop as well as resource allocation (Dębowski et al. 2013; Rajkumar et al. 2014; Ren et al. 2013).

Biomass production of algae is 5–10 times higher as compared to land based plants due to higher photosynthetic effectiveness (Chen et al. 2013). The algae lack lignin, which presents a physical hindrance to hydrolysis by enzyme and cannot be removed by subjecting to pretreatment. This property of algae is useful in the pretreatment and enzymatic hydrolysis in the ethanol process (Sun and Cheng 2002). In an artificial environment, microalgae are grown either in open pond or in photobioreactors

Table 5.3 Number of Operating Second Generation Biorefineries in the World

Region	Pilot	Demonstration	Commercial	Total
Africa	5	0	3	8
Asia-Oceania	6	5	4	15
Europe	7	7	5	19
North America	5	6	9	20
South America	1	1	3	5
Total	24	19	24	67

Based on Nguyen et al. (2017)

(Fig. 5.15). They are also grown in medium rich in nutrients and carbon dioxide. The algal biomass is processed in the same manner as other lipid-based feedstocks for producing biofuel. Bioethanol can also be produced by fermenting the carbohydrates present in the cells.

5.3.1 *Cultivation of Algae*

Growing algae using man-made and natural open ponds is straightforward and cost effective. Laboratory as well as open pond systems are used for growing algae. Different methods which have been used for growing and harvesting algae for ethanol production are summarized in Tables 5.4 and 5.5.

Research conducted earlier on ethanol production from algae shows that they can be grown in the laboratory using different types of growth media and controlled conditions (Takagi et al. 2006; Mandal and Mallick 2009; Mallick et al. 2012; Ren et al. 2013; Feng et al. 2011). Algae are usually grown in beakers, plastic pots and tubs. Several factors affect the production of biomass under controlled conditions. These factors include the following:

- *Intensity and duration of artificial light*
- *Aeration*
- *Concentration of carbon dioxide*
- *Temperature*
- *pH of the growth medium*
- *Supply of nutrients.*

Optimal production of biomass is ensured by controlling these factors. For example, it was found that varying the temperature and pH was found to affect the bioethanol production. *Saccharomyces cerevisiae* IFST-072011 showed the highest ethanol yield at pH of 5.0 to 6.0 at 30 °C (Islam et al. 2012). The low cost sources of nutrients as well as artificial synthetic nutrient medium are used to grow algae for higher production of biomass and growth. For nutrients, nitrogen, phosphorus,

Table 5.4 Cultivation technologies for algal bioethanol biofuel production

Lab cultivation using modified NORO medium amended with 1 M NaCl *Dunaliella* sp. Increased lipid contents 71 (% dry cell weight)
Photobioreactor with cylindrical glass (30 cm length, 7 cm diameter) white fluorescent lights. With varying CO2 sparging conditions *Nannochloropsis oculata* NCTU − 3 Increased lipid 50 (% dry cell weight)
Bioreactor (3.5 L working vlolume) with culture medium BG11 under continuous illumination at an incident irradiance level of 550 μE m − 2 s − 1 and a controlled pH of 6 under carbon dioxide supplemented conditions *Chlorella zofingiensis* Increased lipid contents 55 (% dry cell weight)
Cylindrical glass photobioreactor (30 cm length, 8 cm diameter, light intensity of 70 μmol m − 2 s − 1) with 1 L of working volume at 26 ± 1°C was used under nitrogen starvation and high light intensity Nannochloropsis sp Increased lipid contents 60 (% dry cell weight)
Raceway open ponds each having a capacity of 2000-L was used for algal cultivation in seawater as growth medium, which contained 32 g L − 1 salinity, 0.07 g L − 1 NaNO3, 0.005 g L − 1 NaH2PO4 and traces of FeCl3 *Nannochloropsis sp* Increased lipid contents 52 (% dry cell weight) 48
Algal cultivation was done using paddle wheel-driven open raceway ponds and in a tubular closed photobioreactor (Biocoil) at a salinity of 7% NaCl (w/v) *Tetraselmis* sp. MUR − 233 Increased lipid contents 55 (% dry cell weight) 122
Open solar salt pans using sea water at a temperature of 45 ± 3°C A saline tolerant isolate *Chlorella variabilis* improved the energy output to input ratio minimizing the energy input of 33.12 MJ/kg biomass without affecting its lipid profile 123
Small scale vertical flat-plate photobioreactor (PBR) supplemented with municipal wastewater *Chlorella vulgaris* and *Scenedesmus obliquus* Showed optimal specific growth rates and the TN and TP were completely removed (>99%) from the wastewater within 8 days 125

Based on Takagi et al. (2006), Feng et al. (2011), Moazami et al. (2012), Chiu et al. (2009), Jiang et al. (2011), Raes et al. (2014), Bhattacharya et al. (2016), Yang et al. (2016), Wiley et al. (2009)

and potassium salts or household wastewater, textile wastewater, food industry and municipal wastewater can be used since these waters contain a lot of nutrients.

Growing algae in open ponds either natural environment such as lakes, lagoons, ponds or artificial environment for production of ethanol has been widely studied and is the most commonly used for growing algae (Jorquera et al. 2010; Ashok kumar and Rengasamy 2012). Now a days several companies are exploring this type of growth model for biofuel production. In open pond systems, algae are exposed to natural solar radiation. These systems use shallow (usually one foot deep) ponds. The size ranges from about one acre to several acres. Open pond systems which are mostly

Table 5.5 Harvesting technologies for algal bioethanol biofuel production

Harvesting methods
Dissolved air flotation
Chlorella and *Scenedesmus*
83.7% harvesting efficiency
Electrocoagulation
Chlorella vulgaris
95% harvesting efficiency

Based on Wiley et al. (2009), Vandamme et al. (2011)

used include tanks, shallow big ponds, circular ponds and raceway ponds (Norsker et al. 2011). These are generally made up of a closed loop oval shaped recirculation channels, in general between 0.2 and 0.5 m deep, with blending and spreading which is required for stabilizing the growth of algae and productivity. The open pond cultures are economically attractive, but increase the problems of land use cost, availability of water, lower productivity and suitable climatic conditions. According to one research, "issues related to scale-up algal biomass technology are low cell density and high condensation cost. Further, there is a problem of contamination by fungi, bacteria and protozoa and competition by other microalgae. Moreover, ineffective agitating mechanisms and light limitations are key problems in this system. In spite of the fact that in developed countries commercial cultivation of algae is done, there are weather changes in temperatures and sun light energy throughout the year in most of the regions. Therefore, it is hard to accomplish outdoor mass cultivation of algae year around. However, in tropical developing countries, open cultures of algae can be observed for long period of time in a year because there is neither winter nor cold seasons"(Chen et al. 2011; Patil and Giselrod 2008; Pulz 2001).

Several attempts have been made for making less costly photo bioreactors for growing algae for ethanol production. There are three main types of photo bioreactors (PBRs) (Carvalho et al. 2006):

- *Closed photo bioreactors*
- *Tubular photo bioreactors*
- *Flat-plate photo bioreactors.*

"Closed photo bioreactors can be plates, tubes, or bags. These are made of glass, plastics, or other clear materials, in which the algae are grown with nutrients, light, and carbon dioxide. The closed photo bioreactor could be a technology that removes some of the main issues related with the open pond production systems. However, currently about 100 tons per year, which is 10% of microalgal biomass, is derived from closed photobioreactors, while the major fraction is cultivated in open ponds. One major advantage of closed reactors is the higher yield compared with open systems. It reduces the possibility of external pollution and contamination threats. These systems are more suitable for classified algal strains as the closed shape makes control of probable pollution easier. The more cell mass yield reduces the cutting costs significantly. Only a few of these bioreactors can be practically used for growing algae on a large scale because of high equipment and substrate cost. Although, contamination problem is lesser than open ponds but complete control has not been seen" (Bibi et al. 2017; Carvalho et al. 2006 Moazami et al. 2012; Lehr and Posten 2009; Ugwu et al. 2008; Chen et al. 2011).

Tubular photo bioreactor is preferred for outdoor algal mass cultures because these can provide larger surface area to sun radiations. Mainly outdoor tubular photo bioreactors are made of glass or plastic tube having airlift system or air-pump. The mixing and aeration of the cultures in these bioreactors is performed by airlift systems or air-pump for obtaining higher productivity. These bioreactors can be perpendicular, conical and inclined photo bioreactor. These are relatively inexpensive and have better biomass productivities. The major drawback of tubular photo bioreactors is reduced mass transfer. Also, controlling the temperature of algal culture is difficult. Furthermore, long photo bioreactors may face a problem of pH gradient, dissolved oxygen and carbon dioxide along the tubes. When the pH of the culture increases, the common recarbonation of the cultures result. Some wall growth has been also seen in tubular photo bioreactors (Bibi et al. 2017; Brennan and Owende 2010; Ugwu et al. 2008; Molina et al. 2001; Pirt et al. 1983; Watanabe and Saiki 1997).

Flat plate photo bioreactors contain transparent flat plates for highest capturing of solar energy, and a very thin layer of dense algal culture flows across the plate (Hu et al. 1998; Richmond et al. 2003). Because of this arrangement solar radiation absorbance in the first few millimeters thickness and higher photosynthetic rate is obtained (Guzzon et al. 2008). Such photobioreactors can be used for outdoor mass cultivation of algae having high biomass productivity because of low build up of dissolved oxygen. These reactors can be easily sterilized. Density of photoautotrophic cells is higher than 80 g/l. The photosynthetic efficiency is higher in comparison to tubular versions. The problems are listed below (Brennan and Owende 2010):

- *Difficult temperature control*
- *Some degree of wall growth*
- *Small degree of hydrodynamic stress*
- *Fouling problems.*

Algal biofilms actually play an important role in reducing the major drawbacks of other systems for producing and harvesting algae. The wastewater producing industries are already using biofilms for treatment of wastewater (Wuertz et al. 2003). If sufficient surface area is provided, biofilm growth will be much more effectual and there will be higher production of biomass. An ascendible algal biofilm system can be incorporated into the treatment process, thus increasing the two fold benefit of inexpensive supply of nutrients and treated water. Algal biofilms attached to surface can provide the same more culture density and reduced land and water requirements of immobilized cultures without the related costs of the matrix (US DOE 2010; Cao et al. 2009).

An algal biofilm system includes production, cutting and dewatering steps, perhaps following a more streamlined process with reduced downstream. In this process, the carbohydrates present in the algal biomass are converted into alcohols (ethanol and butanol) by the action of different types of microorganisms. Most extensively used microorganisms are bacteria, yeast and fungi. These microorganisms ferment the carbohydrates into ethanol and carbon dioxide under anaerobic conditions. The carbon dioxide produced in the fermentation process can be used into the ponds for growing microalgae. The fermentation can be presented by the following equation Microalgae has a complex cell wall. Due to this, microorganisms are not able to reach the carbohydrates. Therefore, pretreatment is done. This step releases carbohydrates and converts them into monomers for conversion by microorganisms. Various pretreatment methods have been used. These methods are – physical or mechanical, thermal, chemical and enzymatic. A combined thermal and chemical pre-treatment process is found to be the most commonly used method for releasing sugars. This involves hydrolyzing algal biomass at high temperature in the presence of mild acid or alkali. This pretreatment method gives a high recovery of simple sugars with a higher efficiency because microalgae lack lignin (Mendez et al. 2014). Under certain nutrient starvation conditions (nitrogen and phosphorus), *Chlorella, Chlamydomonas, Scenedesmus, Dunaliella, Tetraselmis* and *Spirulina* possess a higher carbohydrate content (> 40% by weight) (Wang et al. 2015). In green algae and cyanobacteria, the carbohydrates are associated with their cells as storage or structural polysaccharides (www.mdpi.com). Carbohydrates produces glucose and xylose as the major sugars. Other sugars arabinose, mannose and galactose are also produced. Fermenting microorganisms are able to utilize these sugars efficiently (Wang et al. 2015). The most commonly used yeast for production of ethanol on an industrial scale is *Saccharomyces cerevisiae* (www.mdpi.com). This yeast is able to utilize only hexose sugars and produces ethanol, but pentose sugars are not utilized. Higher salt concentration in the hydrolysate of marine microalgae affects the fermentation of *S. cerevisiae*. The biomass of *Arthrospira platensis* was used for ethanol production with a salt stress–adapted *S. cerevisiae* by Markou et al. (2013). Acid treatment and thermal treatments were combined and used; an increase in the acid strength and temperature increased the reducing sugars. Optimal results were obtained by the use of low-strength acids at a higher temperature. High ethanol yield (16.5%) was achieved for *A. platensis* biomass treated with both 0.5 N sulphuric acid- and nitric acid (John et al. 2011). Acidic and enzymatic methods were used for pretreatment of *Chlorella vulgaris* FSP-E. Pretreated biomass was then used for production of ethanol using *Zymomonas mobilis* (Markou et al. 2013). This bacterium produces higher ethanol yield. It does not require controlled addition of oxygen during fermentation. It also shows high utilization of substrate and is able to tolerate higher ethanol concentration up to 120 g/L. The possibility for genetic manipulation is also high (www.mdpi.com). Hydrolysis of *C. vulgaris* FSP-E with enzyme produced about 90.4% glucose and with mild acid hydrolysis about 93.6% glucose was obtained. Use of super high frequency (SHF) with enzymatic hydrolysate produced about 79.9% of the theoretical yield, whereas SHF with acid hydrolysate produced 87.6% of the theoretical yield (Markou et al. 2013; www.mdpi.com). The process for bioethanol production from microalgae is presented in Fig. 5.16 (archive.org).

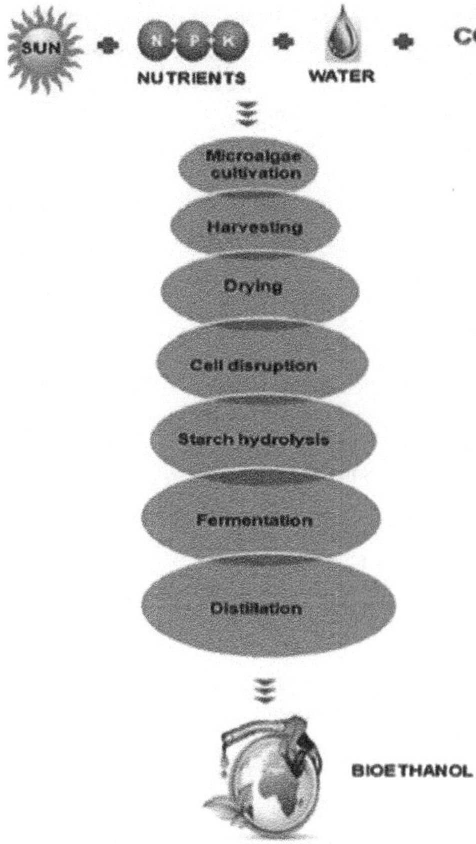

Fig. 5.16 Process for bioethanol production from microalgae. Behera et al. (2015). Reproduced with permission

The worldwide market for ethanol is poised for substantial growth in the next ten years. There is a lot of interest in using algae for biofuel production. Biofuels, petroleum, and agricultural business industries are making huge investments. The leading biofuel producing countries in the world including United States and Europe are not able to grow enough corn, soy, or rapeseed for meeting the biofuels targets (Table 5.6). But, they have the potential for producing algae for producing biofuel on sustainable basis. Therefore, a long term requirement for biofuels in the United States, European Union and Asia will generate new opportunities for algae and other non food-feedstocks for meeting pushy targets for ethanol. For extensive commercialization, several hurdles in production of algal fuels must be conquered. For instance, one of the main hurdles for commercialization of algal bioethanol is the enormous capital cost of facilities. This is essential not only to increase the productivity but also to reduce the vulnerability of microalgae to contaminations. Likewise, some other attributes such as higher lipid content, reduced vulnerability to contamination and tolerance to fluctuations in temperature and salinity are also wanted for cost-effective use of microalgae for efficient production of ethanol. Algal bioethanol also has many

Table 5.6 Commercial production of bioethanol

Company	Investment	Bioethanol
Lanza Tech 2008 Roselle, IL, USA http://www.lanzatech.com	$150 million	100,000 gallons of bioethanol
Algenol 2006 Florida USA http://www.algenol.com	$195 M	8000 gallons per acre per year
Sapphire Energy, Inc 2007 California, US http://www.saphireenergy.com	Greater than $100 million	100,000 gallons/year of fuel-grade ethanol
Joule Unlimited 2007 Bedford, Massachuses, US http://www.jouleunlimited.com	$110million – $70million	
BFS Biopetroleo 2006 Alicante, Spain http://www.biopetroleo.com	Approx.50 M€	42 U.S. gallons
Aurora Algae, Pty 2006 Hayward, California, US http://www.aurorainc.com	2MA$	90 million gallons
BioFields 2006 Mexico City, Mexico www.biofields.com	850 million dollars	946 million litres
Solazyme, Inc 2003 South San Francisco, USA http://www.solazyme.com	More than $160 million	100 million gallons
Aquatic Energy LLC 2007 Lake Charles, LA, USA http://www.AquaticEnergy.com	$32 million	1.5 Million gallon
Abengoa 1941 Seville, SEV, Spain http://www.abengoa.com	92 M€	782 Million gallons
POET LLC 1986 Sioux Falls, SD, USA http://www.poet.com	$8 million	1.7 billion gallons

Based on Bibi et al. (2017)

other limitation in its economic commercialization to public market. One of such limitations is the production cost which then depicts the biofuel price. Capital costs for photobioreactors are very high in comparison to open ponds systems. Nevertheless, photobioreactors can be relocated, reducing the risk of this capital expense.

In open pond systems, liners are very costly. The estimates range from 24 to 75% of capital expenses (Davis et al. 2013; Rogers et al. 2014; Coleman et al. 2014). To reduce the processing cost, the capacity of capital equipment should be designed so that it is not idle for long periods (Abodeely et al. 2014). Freshwater, carbon dioxide and nutrient source also affect the profitability as they can account for up to 30% of production costs for algal biofuels (Chen et al. 2011; Clarens et al. 2010; Zhou et al. 2014). Costs of nutrients can be reduced by using waste streams enriched with carbon dioxide from neighbouring industry or power plants having sufficient amount of nitrogen and phosphorus. Therefore, use of waste water as a source of nutrients may reduce the cost of algal biofuel which is the subject of feasibility analysis (Fortier and

Sturm 2012). Moreover, waste water treatment credits might be redeemed if waste water are used.This would lead energy saving by up to 20% from other sources (Lundquist et al. 2010). To maximize profitability, efficiency in algal biofuel supply chain is quite important. Doubling the biomass productivity might reduce the cost to produce algal biofuel by 26–32% or 40–42%, respectively (Benemann and Oswald 1996; Nagarajan et al. 2013). The strain selection also affects profitableness with different productivities depending on the lipid content and potential for producing co-products. Harvesting and extraction are the main steps contributing the biofuel price (Richardson et al. 2014). Hydro treating at a central conventional refinery would prevent the requirement for hydro treater, hydrogen plant, and import of natural gas (Jones et al. 2014).

Off-take agreements would also affect price and profitability of algal bioethanol (Neste Oil Corporation Neste 2013, 2014). The production of bioethanol using a single source can improve economics of algal biorefineries (Zhu et al. 2014).

The co-products of algal bioenergy are food and feed; therefore on a commercial scale, market saturation could lead to reduced profitability. Furthermore, subsidies and the price of carbon will help determine future profitability of algal biofuels. "It should be concluded that even if the technical barriers are overcome, the practical barriers for instance location, land use, etc. will take several years to overcome and that this will be an evolutionary commercialization, not revolutionary. Environmental problems encounter human resourcefulness and capability to project maintainable resolutions for defending the life of all living organisms including human beings on earth. There are the requirements to protect fresh water and farming lands for manufacturing diet, to struggle the greenhouse effect and to produce energy from non-fossil sources" (Yang et al. 2011a, b; Prasad et al. 2007).

"Fossil hydrocarbons have become scarce and pricey; procedures to change biomass to cheap liquid biofuels are more and more striking. During the recent years, a substantial development in production of biofuels at global level might serve as an initiative to overcome the ecological impacts due to fossil fuels. Production of biofuels from renewable supplies is extensively known to be the most defensible substitutes of petroleum obtained fuels and feasible resources for eco-friendly and economic sustainability. Renewable biofuels play a central part to deal with energy safety, eco-friendliness and the worldwide and nationwide issues of climate change" (Chisti 2007; Cai et al. 2013).

Biofuel prodction from algae is well thoughtout as a novel approach which can be judged in relation to sustainability and ecological preservation. Biofuel production by the use of algae might serve as a potential bioresource as algae can consume large amount of carbon dioxide and produce biomass. Furthermore reducing carbon dioxide, such renewable biofuels offer energy facilities with zero and or almost zero emissions of air contaminants and also greenhouse gases (Bougrier et al. 2006). Algae utilize nutrients such as nitrogen and phosphorous from a variety of waste water sources (example farming run-off, rigorous animal food processes, and manufacturing and community wastewater), and therefore, provide acceptable bio-remediation of these wastewater for environmental and financial assistances (Shilton et al. 2008). Cultivation of algae for biofuels on waste water results in preservation of fresh water resources. This results in preservation of natural water resources, advatageous bioremediation of wastewaters, reduction in emissions and contribution in economical benefits and also energy security (Rodolfi et al. 2009). Biofuels are 'oxygenate' therefore disseminate larger oxygen to petroleum blend, refining the effectiveness of ignition. "Bioethanol reduces the interfacial tension of petrol with reverence to water, which allows the ethanol-gasoline non-aqueous phase liquid to go in minor pore spaces, and to enter more readily through the vadose zone to the water table" (Zhou and Thomson 2009).

"Biofuels are well recognized a reason of dehydration of swelling and also non-swelling clays, making microfractures which increase the clay penetrability. Use of bioethanol also remobilizes sorbed BTEX by suspension. It stops biodegradation of petroleum pollutants, particularly BTEX, by special degradation of the bioethanol. Its biological oxygen demand also navies more reducing situations, which turn into anaerobic and even methanogenic working against the natural attenuation" (Bibi et al. 2017; Ugwu et al. 2008; Ganpathy et al. 2009; Huang et al. 2011; Tian et al. 2009; Thamsiriroj and Murphy 2009).

Production of biofuels using natural bioresources and generated bioenergy makes independent and security of energy supply. Use of agricultural residues and waste substrates as raw materials will reduce the potential clash between food and fuel and also produce the biofertilizers and biopesticides (Singh et al. 2010). Competition of feedstuff supply with food, and pulling down of usual surroundings resulting from plantation of energy crops are some expected matters. Use of biofuels improve quality of air and reduces the generation of harmful air pollutants, greenhouse gases, and acid rain forming sulfur dioxides and play a significant role in reduced ozone diminution (Clarens et al. 2010). Toxicity of bioethanol is low. It decomposes readily. Furthermore, its use generates lesser air-borne contaminants in comparison to petroleum fuel. Bioethanol can substitute octane enhancers such as methylcyclopentadienyl manganese tricarbonyl and aromatic hydrocarbons like benzene or oxygenates like methyl tertiary butyl ether (Champagne 2007).

According to Bibi et al. (2017) "development and use of biofuels as fossil fuels substitute, still needs a further ground breaking technical expansion for proliferating their practicability by increasing the energy equilibrium and reducing the discharges and manufacturing charge, are true substitutes that complete the biofuels forthcoming scheme".

5.4 Production Costs

The production costs of bioethanol is variable (Table 5.7). It depends on the source of biomass. If the production costs for gasoline are 0.25 Euro/L, then this emphasizes the need to have governmental tax rebates in closing the price gap between biofuel and

Table 5.7 Production costs of bioethanol

Biomass source	Production costs €/litre
US corn	0.42
Corn stover	0.45–0.58
EU wheat	0.27–0.43
EU sugarbeet	0.32–0.54
Brazil sugarcane	0.16–0.28
Molasses (China)	0.24
Sweet sorghum (China)	0.22
Corn fibre (US)	0.41
Wheat straw (US)	0.44
Spruce (softwood)	0.44–0.63
Salix (hardwood)	0.48–0.71
Lignocellulose (biowaste)	0.11–0.32
Gasoline	0.25

Based on Sassner et al. (2008), Gnansounou (2008)

fossil fuels. Economic drivers for the production and consumption of all biofuels are linked to the global price of oil. This is obviously a dynamic situation (with increasing oil prices improving the case for biofuels. It is apparent that for first-generation bioethanol feedstocks, Brazilian sugarcane represents one of the cheapest. Goldemberg (2007) estimated the production costs of bioethanol from sugar beet at 26.1 €/GJ, from sugarcane (Brazil) at 7.3 €/GJ; from sugarcane (US and UK) at 12.3 €/GJ, from maize 9.3 €/GJ and from cellulose 20.3 €/GJ. EUBIA (2004) estimated 17.9 €/GJ for bioethanol from sugar beet, 16.7 €/GJ from wheat, 5.2 €/GJ from sugarcane (Brazil) and 13.6 €/GJ from maize. Ecofys (2003) estimated the production costs of bioethanol from straw at 26.0 €/GJ, from sugar beet at 27.9 €/GJ and from wheat at 29.8 €/GJ. Another study (AEA 2003) found 29.3 €/GJ for bioethanol from EU straw or beet pulp, 24.2 €/GJ from EU sugar beet, 21.3 €/GJ from EU wheat, 11.2 €/GJ from US maize, 9.0 €/GJ from Brazilian sugar cane in Brazil, and 31.1 €/GJ from Brazilian sugar cane in the UK. For bioethanol to be economically competitive with fossil fuels, production costs should be no greater than ~0.2€/litre compared with gasoline.

The figures presented in Table 5.7 are approximations due to fluctuating raw material costs. For example, the ethanol production costs in United States are based on $4 bushel corn (32 lbs of starch and 2.8 gals of ethanol). In 2010 cost of sugar cane was at a historical high and current ethanol production costs from this feedstock are estimated at around $0.35 per litre. Lignocellulosic biomass costs are highly feedstock dependent (waste wood and paper costs will vary widely depending on locality and transport costs). Lignocellulose to- ethanol production costs would be expected to become lower in the future as new technology improves the overall conversion processes. Biomass feedstock costs represent the predominant expenditure in bioethanol production, with first generation feedstocks generally 50–80% of total costs, whilst for lignocelluloses bioethanol processes, the feedstock costs are only ~40% of total costs (Petrou and Pappis 2009).

5.5 Important Developments in the Production of Cellulosic Ethanol

Ethanol-from-cellulose (EFC) holds great potential due to the widespread availability, abundance, and relatively low cost of cellulosic materials.Significant investment into research, pilot and demonstration plants is ongoing to develop commercially viable processes utilising the biochemical and thermochemical conversion technologies for ethanol (RFA 2007b). Johnson et al. (2010) reviewed the status of commercial lignocellulosic ethanol production.

According to the International Energy Agency (IEA) database and other sources (Padella et al. 2019; Ernsting and Smolker 2018; Barros and Berk 2018; Kim 2018; Flach et al. 2019; https://demoplants.bioenergy2020.eu/; Schill 2018; Sapp 2018; Lane 2016; Verbio Press Release 2018; Lane 2017; POET-DSM 2018), there are many first-of-a-kind commercial scale cellulosic ethanol plants in the world, although operations of some of the plants are at present idle or on-hold (Table 5.8).

Table 5.9 shows the status of the commercial lignocellulosic ethanol facilities in United States (Zhang 2019). Only a small number of commercial plants are operational in Norway, United States, and Brazil. In addition to the plants shown in Table 5.8, Quad County Corn Processors and six other plants (using the Edeniq technology) converted their conventional corn ethanol refineries to produce ethanol from corn kernel fiber, known as 1.5 generation technology. It will qualify as cellulosic biofuel in United States following the EPA definition. But, corn kernels contain substantial amount of fibers (10–12%) which has an impact on the conversion of fermentable sugar, but they mostly contain starch (Schill 2018); for that reason, they should not be stringently considered as second generation ethanol plants. Other plants have also planned to produce 1.5 generation ethanol. D3Max and ICM are developing their own technologies (Schill 2018).

"Cellulosic ethanol industry is still in its infancy. In the United States, the first commercial scale plants to produce cellulosic biofuels started operating in 2013. In the following 5 years, cellulosic ethanol production grew from 0 to 10 million gallons and most likely topping 15 million in 2018. However, that is far from the Renewable Fuel Standard's original target of 7 billion gallons of cellulosic biofuel by 2018 and 16 billion by 2022. Of all five commercial cellulosic ethanol plants that were built/to be built in the U.S. from 2010 to 2016, only POET's Emmetsburg, Iowa facility is still in operation in 2019. In 2017, the total cellulosic ethanol produced was less than half the nameplate capacity (25 million gallons year − 1) of this single plant"(Table 5.10) (www.forbes.com).

Currently the United States has a target of 136,260 million litres per year (ML/yr) of renewable fuels production by 2022. This target is only achievable with a majority

Table 5.8 Global commercial scale cellulosic ethanol plants

Company	Project	Country output	Capacity (ktons)	Status	Start-up year
Abengoa Bioenergy Biomass of Kansas, LLC	Commercial (acquired by Synata Bio Inc. [21])	US	75	Idle	2014
Aemetis	Aemetis Commercial	US	35	Planned	2019
Beta Renewables (acquired by Versalis [22])	Alpha	US	60	On hold	2018
Beta Renewables (acquired by Versalis)	Energochemica	EU (Slovakia)	55	On hold	2017
Beta Renewables (acquired by Versalis)	Fujiang Bioproject	China	90	On hold	2018
Beta Renewables 1 (acquired by Versalis)	IBP-Italian Bio Fuel	EU (Italy)	40	Idle	2013
Borregaard Industries AS	ChemCell Ethanol	Norway	16	Operational	1938
Clariant	Clariant Romania	EU (Romania)	50	Under construction	2020
COFCO Zhaodong Co	COFCO Commercial	China	50	Planned	2018
DuPont	Commercial facility Iowa (acquired by VERBIO [23])	US	83	Idle	2016
Enviral	Clariant Slovakia	EU (Slovakia)	50	Planned	2021
Fiberight LLC	Commercial Plant	US	18	Under construction	2019
GranBio	Bioflex 1	Brazil	65	Operational	2014
Henan Tianguan Group	Henan 2	China	30	Idle	2011
Ineos Bio	Ineos Bio Indian River County Facility (acquired by Alliance Bio-Products in 2016 [24])	US	24	Idle	NA

(continued)

Table 5.8 (continued)

Company	Project	Country output	Capacity (ktons)	Status	Start-up year
Longlive Bio-technology Co. Ltd.	Longlive	China	60	Idle	2012
Maabjerg Energy Concept Consortium	Flagship integrated biorefinery	EU (Denmark)	50	On hold	2018
POET-DSM Advanced Biofuels	Project Liberty	US	75	Operational	2014
Raízen Energia	Brazil	Brazil	36	Operational	2015
St1 Biofuels Oy in cooperation with North European Bio Tech Oy	Cellunolix®	EU (Finland)	40	Planned	2020

Based on Padella et al. (2019). Reproduced with permission

Table 5.9 The status of the U.S. commercial lignocellulosic ethanol facilities

Company	Location	Raw material	Capacity (mg year^{-1})	Status
DuPont Nevada	Iowa		30	Sold to Verbio in Nov. 2018
Mascoma	Kinross, MI	Wood waste	20	Construction halted in 2013
POET LLC	Emmetsburg, IA	Corn stover	20–25	Operational in Sep. 2014
Abengoa Bioenergy	Hugoton, KS	Wheat straw	25–30	2013–2016 Bankrupt
BlueFire Ethanol	Fulton, MS	Multiple sources	20	Construction halted 2011

Based on Zhang (2019)

of this renewable fuel coming from lignocellulosic material, such as corn stover, wood, switch grass, wheat straw and purpose grown energy crops. Demonstration-scale cellulosic ethanol plants are under construction as part of the government's goal to make cellulosic ethanol cost competitive. The plants cover a wide variety of feedstocks, conversion technologies and plant configurations to help identify viable technologies and processes for full-scale commercialisation. All demonstration plants, which are sized at 10% of a commercial-scale biorefinery, are expected to be operational soon.

Table 5.10 Operating Cellulosic Ethanol Plants in the U.S.

Name of the company	Raw material	Scale	Year of operation	Capacity (mil gal)
American Process, Alpena, MI	Wood chips	Commercial scale	2012	0.95
American Process, Thomaston, GA	Wood chips	Commercial scale	2013	N/A
Calgren Renewable Fuels, Pixley, CA	Cow manure	Commercial scale	2015	N/A
DuPont Nevada, IO	Corn stover	Commercial scale	2015	30
Gulf Coast Energy Livingston, AL	Wood waste	Pilot scale	2009	20
Indian River Bioenergy Center Vero Beach, FL	Municipal solid waste	Commercial scale	2013	8
LanzaTech Soperton, GA	Wood waste	Pilot scale	2014	0.09
Pacific Ethanol Stockton, CA	Corn kernel fiber	Commercial scale	2015	0.75
Project LIBERTY (POET) Emmetsburg, IA	Corn stover, corn cobs, leaves, husk, stalk	Commercial scale	2014	20
Quad-County Galva, IA	Corn kernel fiber, corn	Commercial scale	2014	2
Renmatix Rome, NY	Wood chips, tall grasses, corn stover, bagasse	Demonstration	2008	N/A
Summit Natural Energy Cornelius, OR	Food processing and agricultural waste	Pilot scale	2009	N/A
Tyton Biofuels Raeford, NC	Tobacco waste	Pilot scale	2010	15
ZeaChem Boardman, OR	Wood	Demonstration	2013	0.25

Based on Nguyen et al. (2017), Zhang (2019), Voorhis (2016), Voegele (2014, 2018), McGlashen (2013), POET-DSM (2018)

Demonstration and commercial plants include—Abengoa—Alico, Alltech, American Energy Enterprises (AEE), Bluefire Ethanol, Coskata, Flambeau River Papers, Park Falls, Wisconsin, Fulcrum-Bioenergy, Sierra Biofuels Plant, ICM, Mascoma, The Wisconsin Rapids, Red Shield Environmental (RSE), Pacific Ethanol, The BioGasol process, Poet, Pure Energy & Raven BioFuels, Range Fuels, Verenium, Virent. Several efforts are underway in North America to commercially produce ethanol from wood and other cellulosic materials as a primary product (Yacobucc

2008). Five largest ethanol producers are ADM, Poet, Valero Energy Corporation, Green Plains Renewable Energy, Flint Hills Resources LP https://www.farmprogress.com/ethanol/5-largest-ethanol-producers.

In Brazil, total ethanol production from corn in 2019 was estimated at 1.4 billion liters, an increase of 609 million liters compared to 2018. Total ethanol production from lignocelluloses was estimated at 45 million liters, and represents an insignificant share of total production of ethanol in Brazil. No important changes have been made to the present status of advanced biofuels research, development and production. Total 2019 domestic requirement for ethanol (fuel and other uses) was estimated at 33.93 billion liters, up 2.19 billion liters compared to revised figure for 2018. Ethanol is mostly used for fuel applications in Brazil. Brazil's total ethanol exports are estimated at 1.8 billion liters, an increase of eleven percent compared to total exports in 2018 (1.62 billion liters). Brazil's total ethanol imports for 2019 were projected at 1.2 billion liters, a decrease of 495 million liters relative to the previous year (1.695 billion liters). Ethanol imports are only for fuel use and originate almost entirely from the United States. https://apps.fas.usda.gov/newgainapi/api/report/downloadreportbyfilename?filename=Biofuels%20Annual_Sao%20Paulo%20ATO_Brazil_8-9-2019.pdf.

Canadian government has set it sights on a target of 5% renewable fuel in gasoline and 2% renewable fuel in diesel. To support this, the federal government has established funding of C$550 million dollars for pilot plants and process development and pre-commercial development, with a further C$500 million for demonstration scale facilities and to assist with bridging the gap between development and commercialization of cellulosic ethanol technologies. Facilities under construction or planned include: Iogen, lignol.

Details of facilities under construction or planned in the rest of the world are discussed below.

In South America, Brazil is today producing ~40% of the world's ethanol from sugar cane. Yields vary from 6,600–7,500 L/ha which means production costs half those of US corn-based ethanol processors. Therefore, there has been relatively low interest in second generation bioethanol. However, recent studies made by Brazilian National Development Bank (BNDES) together with FAO show that when fermenting bagasse and sugar cane the ethanol yield could reach 13,000 L/ha. The plants are located in Dedini, Sao Paulo, Brazil and in Chile.

In Europe Abengoa, Spain—Abengoa Bioenergy has been operating a biomass-to-ethanol pilot plant since the end of 2007 at the Biocarburantes Castilla y León grain-ethanol plant in Babilafuente, near Salamanca in Spain. 5 ML/yr are produced from wheat and barley straw using enzymatic hydrolysis (glucose). A steam explosion pretreatment stage, from SunOpta has been installed and is operational. It is intended to separate the lignin and pentose sugars as co-products. Chemrec, Pitea, Sweden—Chemrec has developed a gasification process to convert pulp mill black liquor (BL) to liquid fuels including ethanol. Other plants in Europe are: SEKAB, Ornskoldsvik, Sweden, Stora Enso, Varkaus, Finland UPM Kymmene, Finland, Elsam, Denmark Choren, Freiburg, Germany—DONG Energy, Denmark, BioGasol Denmark, TMO Renewables, Guildford, UK In Japan, Australasia, Asia; following plants are under construction or planned. The BioEthanol Japan plant in Osaka Prefecture; Honda-RITE, Japan; Marubeni, Saraburi, Thailand; China Resources Alcohol Corporation; China; Mission NewEnergy, India; Ethtec, Australia—Ethtec, a Willmott Forests subsidiary; Pure Power, Singapore; LanzaTech, New Zealand; Scion, Rotorua, New Zealand.

A University of Florida researcher has developed a biotech "bug" that is capable of converting cellulosic biomass to ethanol. Lonnie Ingram, Director of the Florida Center for Renewable Chemicals and Fuels, has developed genetically engineered *E. coli* bacteria that can convert all types of sugar found in plant cell walls into fuel ethanol (Newswise 2005). The bacteria produce a high yield of ethanol from biomass such as sugarcane residues, rice hulls, forestry and wood wastes, and other organic materials. Ingram says he genetically engineered the *E. coli* organisms by cloning the unique genes needed to direct the digestion of sugars into ethanol, the same pathway found in yeast and higher plants. With the ethanol genes, he says that bacteria produce ethanol from biomass sugars with 90–95% efficiency. Ingram began research in this area in 1985 and says that Dr. Nancy Ho of Purdue University has also made considerable progress in engineering yeasts to use in this process. "Reducing the cost and improving the efficiency of converting cellulosic materials into fermentable sugars is one of the keys to progress. The Department of Energy's National Renewable Energy Laboratory (NREL) has partnered with private biotech companies to make significant strides in this area." Ingram's University of Florida technology has become Landmark Patent No. 5,000,000 through the U.S. Department of Commerce. It is being commercialized with assistance from the Department of Energy, and BC International Corp., based in Dedham, Massachusetts, holds exclusive rights to use and license the engineered bacteria.

NREL and its partners say that the research conducted in this area is an important step toward realizing the potential of biorefineries (www.ethanol.org/documents/6-05_Cellulosic_Ethanol.pdf). Biorefineries, analogous to today's oil refineries, will use plant and waste materials to produce an array of fuels and chemicals—not just ethanol. Biorefineries will extend the value-added chain beyond the production of renewable fuel only. Progress towards a commercially viable biorefinery depends on the development of real-world processes for biomass conversion. With these new technologies for the production of cellulosic ethanol, its promise becomes closer to reality with each passing day.

The Department of Energy (DOE) announced that six companies—Abengoa Bioenergy Biomass of Kansas, LLC ALICO, Inc. BlueFire Ethanol, Inc. Broin Companies, Iogen Biorefinery Partners, LLC, Range Fuels—have been awarded cellulosic ethanol grants to help with the construction of cellulosic ethanol biorefineries. The release of these grants is a major step forward for the cellulosic ethanol industry as many ethanol producers were awaiting this announcement to move forward with their plans. Renewable Fuels Association President said that "Cellulosic ethanol production is a must if we truly aspire to move this country toward a more diverse energy future. While corn will remain a key component of our ethanol industry, the kind of production necessary to greatly reduce gasoline consumption in this country can only be realized from the addition of cellulosic material as a feedstock. These grants are critical to bringing cellulosic ethanol to the commercial market and underscore the important partnership the federal government must have with the U.S. ethanol industry to achieve both our short-term and long-term energy goals." The grants are designed to help ethanol producers with the upfront capital costs associated with construction of cellulosic ethanol biorefineries. The stated DOE goal is to prove the feasibility of cellulosic ethanol technology. Recipients are eligible for up to $100 million and must show a 60% industry/40% government cost share. In addition, DOE is working on a loan guarantee program for cellulosic ethanol biorefineries. This program has seen significantly slower progress but is equally as important to helping companies construct cellulosic ethanol facilities often costing 4-to-5 times that of a traditional corn ethanol biorefinery.

Cellulosic ethanol is on track to be cost competitive with corn-based ethanol, a development that could drive the fuel's production, according to an industry survey conducted by Bloomberg New Energy Finance. The survey focused on 11 major players in the cellulosic ethanol industry, all of which use a technique known as enzymatic hydrolysis to break down and convert the complex sugars in non-food crop matter, and a fermentation stage to turn the material into ethanol, BNEF said. Cellulosic ethanol cost 94 cents a liter to produce in 2012, about 40% more than ethanol made from corn, BNEF said. "That price gap will close in the near future, surveyed cellulosic ethanol producers predicted. Project capital expenditures, feedstock and enzymes used in the production process are still the largest costs of running a cellulosic ethanol plant, the respondents said in the survey. But technology has pushed operating costs lower. For example, enzyme costs for a liter of cellulosic ethanol dropped 72% between 2008 and 2012 due to technological improvements, BNEF said. Cellulosic ethanol producers will shift their focus from technology enhancements to logistical planning over the next five to 10 years in an effort to rein in capital costs, suggesting the industry is maturing, said BNEF's lead biofuel analyst Harry Boyle. Globally, there are 14 enzymatic hydrolysis pilots, nine demonstration-stage projects and 10 semi-commercial scale plants either announced, commissioned or due online shortly, according to the survey. Five of the semi-commercial plants are in the US and more are expected to open in Brazil in the near future, BNEF said. A semi-commercial facility with a capacity of 90 million liters per year requires an initial capital outlay of about $290 million" (www.environmentalleader.com). With commissioning of second- and third-generation plants with capacities between 90 m and 125 m liters, initial capital costs per installed liter are expected to fall from $3 to $2 due to economies of scale and a reduction in over-engineering, BNEF said.

5.5.1 Energy Balances

Bioethanol produced from lignocellulosic biomass and other biowaste materials generally result in very favourable (i.e. positive) NER (net energy ratio) values. The definition of net energy value (NEV) is the difference between the energy in the fuel product (output energy) and the energy needed to produce the product (input

Table 5.11 Energy balances for bioethanol production from different feedstocks

Feedstock	Energy balance
Maize	1–2
Sugar cane	6.5–9.5
Sugar beet	1.1–2.3
Sweet sorghum	0.9–1.1
Lignocellulose	Highly dependent on feedstock, but generally highly positive
Gasoline	6

Based on Walker (2010)

energy). Ethanol has a net positive energy balance. It takes less than 35,000 BTUs of energy to turn corn into ethanol, while the ethanol offers at least 77,000 BTUs of energy which shows that ethanol's energy balance is clearly positive (Shapouri et al. 1995, 2002, 2003; Shapouri and McAloon 2004; Lorenz and Morris 1995; Wang et al. 1999; Kim and Dale 2004a, b; Farrell et al. 2006) and has an extremely high petroleum/fossil energy displacement ratio.

A similar useful parameter in this regard is the Net Energy Balance (NEB), which is the ratio of the ethanol energy produced to the total energy consumed (in biomass growth, processing and biofuel production). Table 5.11 presents energy balances from the production of bioethanol from sugarcane, maize and lignocellulose, and it is apparent that of the first-generation biomass sources, sugar cane represents the most favourable feedstock with respect to energy balance. Energy balance values <1 mean that bioethanol production is unfeasible from energetic standpoint and is indicative of excess of fossil energy used to produce bioethanol. For maize (corn) ethanol processes in North America (USA), typical values are 1–2:1, whilst for sugarcane ethanol processes in South America (Brazil) typical values are 5–10:1. Figures are variable due to different geographic, climatic and agricultural reasons, but for Brazilian sugarcane ethanol operations, a typical energy balance of 8 (i.e. 8 times energy production in comparison to inputs) and GHG reductions of 90% (compared to only 30% for ethanol from corn) are achievable (Amorim et al. 2009; Basso and Rosa 2010). Brazilian bioethanol plants that combust residual bagasse to steam for electricity generation have very favourable energy balances. Brazil is thus considered to be a sustainable biofuel producer. Calculations of energy balances in bioethanol production depend on several factors, for example, whether or not fossil fuel usage in agronomic practices and co-generation of energy from by-products are included. Nevertheless, there is scope to reduce energy inputs from the bioprocessing (rather than biomass cultivation) perspective, particularly through adoption of modern biotechnology (Mousdale 2008).

The advances in the farming community as well as technological advances in the production of ethanol have led to positive returns in the energy balance of ethanol. Studies have shown that the ethanol energy balance is improving by the year (Wang 2005; Shapouri et al. 1995, 2002, 2003; Lorenz and Morris 1995; Wang 1999; Wang et al. 1999; Morris 1995). The energy output to energy input ratio for converting irrigated corn to ethanol is now 1.67–1. According to USDA report, 1995, the corn ethanol energy balance had a gain of 24%. That same report was revisited the next year, the authors concluded the ratio had risen to 34%. This number is reinforced by a 2002 report. The report concluded that ethanol production is energy efficient because it yields 34% more energy than is used. The USDA in June 2004, again looked at this issue and determined that ethanol continues to be more efficient, and now provides the aforementioned 1.67–1 gain in energy. Many advances have led to the surge in ethanol production efficiency. One key issue is the ability to produce more gallons of ethanol per bushel of corn. In the early 1990's, plants were able to produce about 2.5 gallons of ethanol per bushel. That number has since increased to between 2.7 and 2.8 gallons per bushel today.

Crop with a higher sugar content than corn, such as sugar beets, would result in production with a much higher positive net energy balance (www1.eere.energy.gov/biomass/net_energy_balance.html). If corn farmers use state-of-the-art, energy efficient farming techniques, and ethanol plants use state-of-the-art production processes, then the amount of energy contained in a gallon of ethanol and the other co-products is more than twice the energy used to grow the corn and convert it into ethanol. These studies indicated an industry average net energy gain of 1.38–1. The industry-best existing production net energy ratio was 2.09–1. If farmers and industry were to use all the best technologies and practices the net energy ratio would be 2.51–1. In other words, the production of ethanol would result in more than 2-1/2 times the available energy than it took to produce it. A 1999 study by Argonne National Laboratory found the energy balance of cellulosic ethanol to be in excess of 60,000 Btu per gallon (Wang 1999). Given that feedstocks for cellulosic ethanol are essentially waste products like corn stover, rice bagasse, forest thinnings or even municipal waste, there are relatively few chemical and energy inputs that go into the farming of feedstocks for cellulosic ethanol. A secondary factor, although to a much lesser extent, is the fact that cellulosic ethanol plants will presumably produce "extra" energy that can be fed into the power grid. Doing so will effectively displace the use of electricity produced in power plants, which for the most part rely upon fossil fuels. Ethanol's net energy value published by different researchers (Shapouri et al. 1995, 2002, 2003; Lorenz and Morris 1995; Wang et al. 1999; Pimental 2002) is shown in Table 5.12.

Cellulosic ethanol from agricultural feedstocks and energy crops is usually considered to be environmentally sustainable because it provides higher reduction of GHG emissions and zero or low indirect emissions from land use change in comparison to traditional ethanol production from food and feed crops. Focusing on the framework after 2020, the RED recast (European Parliament and Council of the European Union 2018) states that biofuels have to meet a 65% greenhouse gas reduction threshold, in comparison to fossil fuels, in plants starting operation from 1 January 2021. Table 5.13 presents estimated greenhouse gas emissions associated with bioethanol production from cereal straws and perennial crops.

Several sources report costs as the major issue for the advancement of the cellulosic ethanol (Badger 2002; Kumar et al. 2016; Liu et al. 2019). Results from several studies exploring production costs of advanced biofuels, including cellulosic ethanol,

Table 5.12 Ethanol's Net Energy Value reported by different researchers

NEV (Btu)
+20,436 (HHV)
+30,589 (HHV)
+29,826 (LHV)
+22,500 (LHV)
−33,562 (LHV)
+21,105 (HHV)
+23,866 to +35,463 (LHV)
+17,508
−22,300
+21,105
+30,258 (LHV)

Shapouri et al. (1995, 2002, 2003), Lorenz and Morris (1995), Wang et al. (1999), Pimental (2002)

Table 5.13 Estimated greenhouse gas (GHG) emissions (direct and indirect) associated with cellulosic ethanol production from cereal straw and perennial crops

	Direct Emissions from RED Recast, Annex V—Default Values (g CO_2 eq/MJ)	Indirect emissions (ILUC Emissions) (g CO_2 eq/MJ)	Indirect emissions (Displacement Emissions) (g CO_2eq/MJ)
Ethanol from wheat straw	16	16 if unsustainable straw removal is assumed (more than 33–50%) 0 if sustainable straw removal is assumed (less than 33–50%)	8 displaced uses: Livestock bedding and feed; mushroom cultivation; horticulture; heat and power
Ethanol from perennial grass (miscanthus and swithgrass)	16	−12	0 no existing uses

European Parliament and Council of the European Union (2018) Searle et al. (2017), Valin et al. (2015)

are presented below. These studies are: the Joint Research Centre (JRC), Eucar and Concawe (JEC) cost analysis (Padella et al. 2018), the 2017 European Commission report (2017) by the Sub Group on Advanced Biofuels (SGAB) and the 2016 report from the International Renewable Energy Agency (IRENA) (Alberts et al. 2016). For bioethanol from agricultural residues or woody biomass, cost ranges reported by the three studies are presented in Fig. 5.17 (Alberts et al. 2016; Padella et al. 2019;

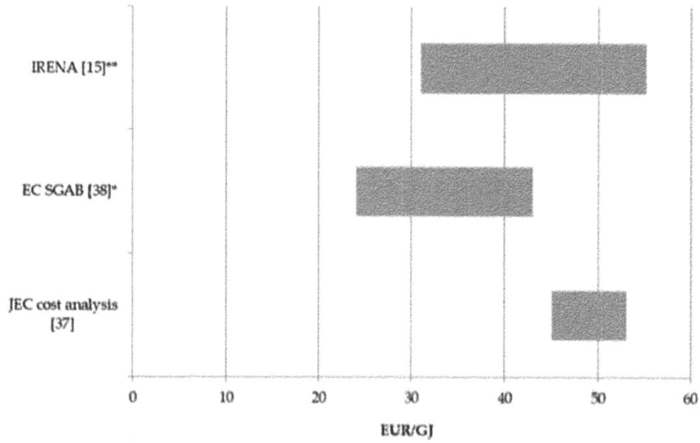

Fig. 5.17 Production costs ranges of cellulosic ethanol from different studies. Padella et al. (2019). Reproduced with permission

European Commission 2017). The cost is dependent upon the best guess on the several components of the cost calculations and, especially the cost of raw material, capital cost and the size of the plants. Cellulosic ethanol is facing major technology challenges as it is in early stage of commercialization and is showing substantial operational and capital costs because of the intricacy of the conversion processes and low maturity of the technology, but there is a possibility for cost reduction in the future (Padella et al. 2018). The cost of raw material is the main contributing factor depending on its accessibility, cost of transportation, and also the different use. Raw materials which are used in cellulosic ethanol plants have generally regional prices and they are traded locally. Their price depends on the local amount of production and also their competing uses (Chudziak et al. 2017). Enzyme cost also represents a relevant cost component of the operational costs and reducing this cost is key to producing cellulosic ethanol economically viable (Alberts et al. 2016; Padella et al. 2018; Johnson 2016). Comparing the costs of three strategies for producing enzymes (off-site, on-site, and integrated), it was found that the integrated method has the lowest cost. Usually, the reported production costs for cellulosic ethanol appear to be significantly higher in comparison to price of conventional ethanol and far from being competitive with fossil fuel (Chudziak et al. 2017).

USDA made an analysis of the energy balance of corn ethanol produced at dry mill biorefineries (Gallagher et al. 2016). It has been concluded that energy balance ratio of corn ethanol today is "between 2.1 and 2.3," describing the result as a small but positive improvement since 2008. The USDA report also analyzed a model dry mill in Iowa that is marketing wet distillers grains and is more closely situated to feedstock supplies and ethanol markets, therby reducing energy use versus the reference case. Ethanol from the model Iowa biorefinery configuration showed an energy balance of 4.0.

USDA's earlier report on energy balance of corn ethanol (Shapouri et al. 2010), which was based on 2008 data, concluded that the energy balance ratio at that time was between 1.9 and 2.3. The new report suggests the average energy balance of corn ethanol is essentially the same today as it was in 2008. This result seems at odds with observed reductions in energy use by dry mill biorefineries since 2008 (Wang et al. 2012; Mueller and Kwik 2013; Christianson and Associates 2016). Several dry mills using more proficient energy systems completed construction after 2008, and many older mills have undertaken retrofits and used new technologies. These developments have resulted in substantially lower average energy use at dry mills today than in 2008.This study was undertaken by RFA for two reasons: "(1) to understand the apparent lack of improvement in corn ethanol's energy balance since 2008 suggested by USDA's latest analysis; and (2) to examine the impact of using more current biorefinery energy use data on the energy balance ratio" (ethanolrfa.org). Two primary explanations for the lower-than-expected energy balance ratio reported in USDA's latest analysis can be found.

"The new USDA analysis relies on the same 2008 survey of dry mill energy use that was used for the previous analysis published in 2010. The new USDA analysis assumes for its reference case that all distillers grains require drying, which increases energy use. Further, impacts of corn distillers oil removal are not considered. When more current and robust data regarding dry mill biorefinery energy use are considered, it was found that the average energy balance of corn ethanol is more likely in the range of **2.6–2.8**, with the top quartile of dry mill biorefineries averaging **3.2–3.4** and some plants very likely achieving the **4.0** energy balance in USDA's *Iowa Wet DG* case" (RFA 2016).

Energy assessment of second generation ethanol production from wheat straw in Indian scenario was studied by Mishra et al. (2018). "Energy efficiency was determined by calculating Net energy ratio (NER), i.e. ratio of output energy obtained by ethanol and input energy used in ethanol production. Energy consumption and generation at each step was calculated briefly (11,837.35 MJ/ha during Indian dwarf irrigated variety of wheat crop production and 7.1148 MJ/kg straw during ethanol production stage). Total energy consumption was calculated as 8.2988 MJ/kg straw whereas energy generation from ethanol is 15.082 MJ/kg straw; resulting into NER > 1. Major portion of agricultural energy input was contributed by diesel and fertilisers whereas refining process of wheat straw feedstock to ethanol and by-products require mainly in the form of steam and electricity. On an average, 1671.8 kg water free ethanol, 930 kg lignin rich biomass (for combustion), and 561 kg C5-molasses (for fodder) per hectare are produced. Findings of this study, net energy ratio (1.81) and figure of merit (14.8028 MJ/nil kg carbon) proves wheat straw as highest energy efficient lignocellulosic feedstock for the country" (https://www.worldwidescienc e.org).

References

Abbas C (2010) Going against the grain: food versus fuel uses of cereals. In: Walker GM, Hughes PS (eds) Distilled spirits- new horizons: energy, environment and enlightenment. Proceedings of the worldwide distilled spirits conference, Edinburgh, 2008. Nottingham University Press

Abbi M, Kuhad RC, Singh A (1996) Bioconversion of pentose sugars to ethanol by free and immobilized cells of *Candida shehatae* NCL-3501: fermentation behaviour. Proc Biochem 31:55–560

Abodeely JM, Coleman AM, Stevens DM, Ray AE, Cafferty KG, Newby DT (2014) Assessment of algal farm designs using a dynamic modular approach. Algal Res 2014(5):264–273

AEA (2003) International resource costs of biodiesel and bioethanol. AEA Technology in commission of the UK Department of Transport. http://www.dft.gov.uk/pgr/roads/environment/research/cqvcf/internationalresourcecostsof3833?page=3

Alberts G, Ayuso M, Bauen A, Boshell F, Chudziak, C, Gebauer JP, German L, Kaltschmitt M, Nattrass L, Ripken R (2016) Innovation Outlook, Advanced Liquid Biofuels; International Renewable Energy Agency (IRENA): Abu Dhabi, UAE, 2016. https://www.irena.org/-/media/Files/IRENA/Agency/Publication/2016/IRENA_Innovation_Outlook_Advanced_Liquid_Biofuels_2016.pdf

Alkasrawi A, Rudolf A, Lid′en G, Zacchi G (2006) Influence of strain and cultivation procedure on the performance of simultaneous saccharification and fermentation of steam pretreated spruce. Enzyme Microb Technol 38:279–286

Amoah J, Ishizue N, Ishizaki M, Yasuda M, Takahashi K, Ninomiya K, Yamada R, Kondo A, Ogino C (2017) Development and evaluation of consolidated bioprocessing yeast for ethanol production from ionic liquid-pretreated bagasse. Bioresour Technol 245:1413–1420

Amorim HV, Basso LC and Lopes ML (2009) Sugar cane juice and molasses, beet molasses and sweet sorghum: composition and usage. In: The alcohol textbook, 5th edn. Nottingham University Press, pp 39–46

Ashok kumar V, Rengasamy R (2012) Mass culture of Botryococcus braunii Kutz under open raceway pond for biofuel production. Bioresour Technol 104:394–399

Badger PC (2002). Ethanol from cellulose: a general review. In: Janick J, Whipkey A (eds) Trends in new crops and new uses. ASHS Press, Alexandria, VA, USA, pp 17–21

Bajpai P, Margaritis A (1982) Direct fermentation of D-xylose to Ethanol by *Kluyveromyces marxianu*. Appl Environ Microbiol 44(5):1039–1041

Bajwa PK, Pinel, D Martin VJJ, Trevors JT, Lee H (2010) Strain improvement of the pentose-fermenting yeast Pichia stipitis by genome shuffling. J Microbiol Methods 81:179–186

Bajwa PK, Shireen T, D'Aoust F, Pinel D, Martin VJJ, Trevors JT, Lee H (2009) Mutants of the pentosefermenting yeast Pichia stipitis with improved tolerance to inhibitors in hardwood spent sulfite liquor. Biotechnol Bioeng 104:892–900

Banat IM, Nigam P, Singh D, Marchant P, McHale AP (1998) Ethanol production at elevated temperatures and alcohol concentrations. Part I: yeasts in general. World J Microbiol Biotechnol 14:809–821

Banerjee S, Mudliar S, Sen R, Giri B, Satpute D, Chakrabarti T, Pandey RA (2010) Commercializing lignocellulosic bioethanol: technologybottlenecks and possible remedies. Biofuels, Bioprod Biorefin 4:77–93

Barros S, Berk C (2018) Brazil Biofuels annual 2018; global agricultural information network (GAIN) report number: BR18017; US Department of Agriculture (USDA) Foreign Agricultural Service: Washington, DC, USA

Basso LC and Rosa CA (2010) Sugar cane for potable and fuel ethanol. In: Proceedings of the worldwide distilled spirits conference, Edinburgh, 2008. Nottingham University Press (in press)

Bauer NA, Gibbons WR (2012) Saccharification versus simultaneous saccharification and fermentation of kraft pulp. Biol Eng 2012(5):48–55

Behera S, Singh R, Arora R, Sharma NK, Shukla M, Kumar S (2015) Scope of algae as third generation biofuels, Frontiers in bioengineering and biotechnology. Mar Biotechnol 90(2):1–13

Benemann JR, Oswald WJ (1996) Systems and economic analysis of microalgae ponds for conversion of CO {sub 2} to biomass. California University, Berkeley, CA (United States). Department of Civil Engineering, [Final report (No. DOE/PC/93204–T5]

Bhattacharya S, Maurya R, Mishra SK, Ghosh T, Patidar SK, Paliwal C, Chokshi K, Pancha I, Maiti S, Mishra S (2016) Solar driven mass cultivation and the extraction of lipids from Chlorella variabilis: a case study. Algal Res 14:137–142

Bibi R, Ahmad Z, Imran M, Hussain S, Ditta A, Mahmood S, Khalid A (2017) Algal bioethanol production technology: a trend towards sustainable development. Renew Sustain Energy Rev 71:976–985

Binod P, Sindhu R, Singhania RR, Vikram S, Devi L, Na-galakshmi S, Kurien N, Sukuraman RK, Pandey A (2010) Bioethanol production from rice straw: an overview. Bioresour Technol 101(13):4767–4774

Björling T, Lindman B (1989) Evaluation of xylose-fermenting yeasts for ethanol production from spent sulfite liquor. Enzyme Microb Technol 11:240–246

Boshoff S, Gottumukkala LD, van Rensburg E, Görgens J (2016) Paper sludge (PS) to bioethanol: evaluation of virgin and recycle mill sludge for low enzyme, high-solids fermentation. Bioresour Technol 203:103–111

Bougrier C, Delgenes JP, Carrere H (2006) Combination of thermal treatments and anaerobic digestion to reduce sewage sludge quantity and improve biogas yield. Process Saf Environ Prot 84(4):280–284

Branco RHR, Serafim LS, Xavier AMRB (2019) Second Generation Bioethanol Production: on the use of Pulp and paper industry wastes as feedstock. Fermentation 5:4. https://doi.org/10.3390/fermentation5010004

Brennan L, Owende P (2010) Biofuels from microalgae, a review of technologies for production, processing, and extractions of biofuels and co-products. Renew Sustain Energy Rev 2010(14):557–577

Bušic A, Marđetko N, Kundas S, Morzak G, Belskaya H, Ivancic Santek M, Komes S, Novak S, Santek B (2018) Bioethanol production from renewable raw materials and its separation and purification: a review. Food Technol Biotechnol 56:289–311

Buzała KP, Kalinowska H, Małachowska E, Przybysz P (2017) The utility of selected kraft hardwood and softwood pulps for fuel ethanol production. Ind Crop Prod 108:824–830

Cai F, Axen SD, Kerfeld CA (2013) Evidence for the widespread distribution of CRISPRCas system in the phylum cyanobacteria. RNA Biol 2013(10):687–693

Cao J, Yuan W, Pei ZJ, Davis T, Cui Y, Beltran M (2009) A preliminary study of the effect of surface texture on algae cell attachment for a mechanical–biological energy manufacturing system. J Manuf Sci Eng 131:1–4

Carvalho AP, Meireles LA, Malcata FX (2006) Microalgal reactors: a review of enclosed system designs and performances. Biotechnol Prog 22:1490–1506

Champagne P (2007) Feasibility of producing bio-ethanol from waste residues: a Canadian perspective feasibility of producing bio-ethanol from waste residues in Canada. Resour Conserv Recycl 2007(50):211–230

Chandel AK, Chan ES, Rudravaram R, Lakshmi M, Venkateswar R, Ravindra P (2007) Economics and environmental impact of bioethanol production technologies: an appraisal. Biotechnol Mole Boil Rev 2(1):014–032

Chen CY, Yeh KL, Aisyah R, Lee DJ, Chang JS (2011) Cultivation, photobioreactor design and harvesting of microalgae for biodiesel production: a critical review. Bioresour Technol 102:71–81

Chen CY, Zhao XQ, Yen HW, Ho SH, Cheng CL, Lee DJ, Bai FW, Chang J (2013) Microalgae-based carbohydrates for biofuel production. Biochem Eng J 78:1–10. https://doi.org/10.1016/j.bej.2013.03.006

Chen P, Min M, Chen Y, Wang L, Li Y, Chen Q, Wang C, Wan Y, Wang X, Cheng Y, Deng S, Hennessy K, Lin X, Liu Y, Wang Y, Martinez B, Ruan R (2009) Review of the biological and engineering aspects of algae to fuels approach. Int J Agric Biol Eng 2(4):1–30

Chisti Y (2007) Biodiesel from microalgae. Biotechnol Adv 2007(25):294–306

Chiu SY, Kao CY, Tsai MT, Ong SC, Chen CH, Lin CS (2009) Lipid accumulation and CO2 utilization of Nannochloropsis oculata in response to CO2 aeration. Bioresour Technol 100:833–838

Chudziak C, Alberts G and Bauen A (2017) Ramp Up of Lignocellulosic Ethanol in Europe to 2030. In: Proceedings of the E4Tech report for the Co-Sponsors of the 6th international conference on Lignocellulosic Ethanol (BetaRenewables DuPont, ePURE, Leaf, Novozymes, Shell St1), Brussels, Belgium, 27–28 Sept 2017

Clarens AF, Resurreccion EP, White MA, Colosi LM (2010) Environmental life cycle comparison of algae to other bioenergy feedstocks. Environ Sci Technol 44(5):1813–1819

Coleman AM, Abodeely JM, Skaggs RL, Moeglein WA, Newby DT, Venteris ER, Vigmosta MS (2014) An integrated assessment of location-dependent scaling for microalgae biofuel production facilities. Algal Res 5:79–94

Davis R, Biddy M and Jones S (2013) Algal lipid extraction and upgrading to hydrocarbons technology pathway. Golden, CO, and Richland, WA: National Renewable Energy Laboratory and Pacific Northwest National Laboratory; 2013, [NREL/TP-5100-58049 and PNNL-22315]

Dębowski M, Zieliński M, Grala A, Dudek M (2013) Algae biomass as an alternative substrate in biogas production technologies—review. Renew Sustain Energy Rev 27:596–604

De Wever, Leliaert Verleyen AF, Vanormelingen E, Van der Gucht P, Hodgson KDA, Sabbe K, Vyverman W (2009) Hidden levels of phylodiversity in Antarctic green algae: further evidence for the existence of glacial refugia. Proc R Soc B: Biol Sci 276:3591–3599

Dias MOS, Maciel Filho R, Mantelatto PE, Cavalett O, Rossell CEV, Bonomi A, Leal MRLV (2015) Sugarcane processing for ethanol and sugar in Brazil. Environ Dev 15:35–51. https://doi.org/10.1016/j.envdev.2015.03.004

Dien BS, Cotta MA, Jeffries TW (2003) Bacteria engineered for fuel ethanol production current status. Appl Microbiol Biotechnol 63:258–266

Dien BS, Jung HJG, Vogel KP, Casler MD, Lamb JAFS, Iten L, Mitchell RB, Sarath G (2006) Chemical composition and response to dilute acid pretreatment and enzymatic saccharification of alfalfa, reed canary grass and switch grass. Biomass Bioenergy 30(10):880–891

Dien BS, Nichols NN, O'Bryan PJ, Bothast RJ (2000) Development of new ethanologenic *Escherichia coli* strains for fermentation of lignocellulosic biomass. Appl Biochem Biotechnol 84(86):181–196

Doan QC, Moheimani NR, Mastrangelo AJ, Lewis DM (2012) Microalgal biomass for bioethanol fermentation: implications for hypersaline systems with an industrial focus. Biomass Bioenergy 46:79–88. https://doi.org/10.1016/j.biombioe.2012.08.022

Ebringerova A, Hromadkova Z (2002) Effect of ultrasound on the extractability of corn bran hemicelluloses. Ultrason Sonochem 9(4):225–229

Ecofys (2003) Biofuels in the Dutch market: a fact-finding study, Utrecht, the Netherlands, report no. 2GAVE-03.12

Edgardo A, Carolina P, Manuel R, Juanita F, Baeza J (2008) Selection of thermotolerant yeast strains Saccharomyces cerevisiae for bioethanol production. Enzyme Microb Technol 43:120–123

Ek M, Gellerstedt G, Henriksson G (2009) Pulp and paper chemistry and technology, vol 1, 1st edn. De Gruyter, Berlin, Germany, pp 1–3, 46, 103, 182, 184, 188. ISBN 978-3-11-021339-3

Ernsting A, Smolker R (2018) Dead end road: the false promise of cellulosic biofuels. Biofuelwatch Report. https://www.biofuelwatch.org.uk/2018/dead-end-road/

European Commission (2017) Sub Group on Advanced Biofuels (SGAB). Sustainable Transport Forum. Building Up the Future, Final Report; European Commission, Brussels, Belgium

European Parliament and Council of the European Union (2018) Directive 2018/2001 of the European Parliament and of the Council of 11 Dec 2018 on the Promotion of the Use of Energy from Renewable Sources (recast). European Parliament and Council of the European Union, Brussels, Belgium

Evtuguin DV (2016) Sulphite pulping. In: Belgacem MN, Pizzi A (eds.) Lignocellulosic fibers and wood handbook, 1st edn. Wiley, Hoboken, NJ, USA, pp. 225–229, 233, 239, 240, 243, ISBN 978-1-118-77352-9

Fan LT, Gharpuray MM, Lee YH (1987) Cellulose hydrolysis biotechnology monographs. Springer, Berlin, p 57

Farrell AE, Plevin RJ, Turner BT, Jones AD, Hare MO, Kammen DM (2006) Ethanol can contribute to energy and environmental goals. Science 311 (Jan 27)

Feng D, Chen Z, Xue S, Zhang W (2011) Increased lipid production of the marine oleaginous microalgae Isochrysis zhangjiangensis (Chrysophyta) by nitrogen supplement. Bioresour Technol 2011(102):6710–6716

Ferguson A (2003) Implication of the USDA 2002 update on ethanol from corn. The Optim Popul Trust, Manch, UK 3(1)

Figueira D, Cavalheiro J, Ferreira B (2017) Purification of polymer-grade fumaric acid from fermented spent sulfite liquor. Fermentation 2017(3):13

Flach B, Lieberz S, Bolla S (2019) EU Biofuels Annual 2019; Global Agricultural Information Network (GAIN) Report Number: NL1902; US Department of Agriculture (USDA) Foreign Agricultural Service: Washington, DC, USA

Fornell R, Berntsson T (2012) Process integration study of a kraft pulp mill converted to an ethanol production plant—Part A: potential for heat integration of thermal separation units. Appl Therm Eng 35:81–90

Fornell R, Berntsson T, Åsblad A (2012) Process integration study of a kraft pulp mill converted to an ethanol production plant—Part B: techno-economic analysis. Appl Therm Eng 42:179–190

Fortier MOP and Sturm BSM (2012) Geographic analysis of the feasibility of collocating algal biomass production with wastewater treatment plants. Environ Sci Technol 2012(46):11426–11434

Galbe M, Zacchi G (1992) Simulation of ethanol production processes based on enzymatic hydrolysis of lignocellulosic materials using Aspen. Appl Biochem Biotechnol 34–35:93–104

Ganpathy TK, Murugesan, Gakkar RP (2009) Performance optimization of Jatropha biodiesel model using Taguchi approach. Appl Energi 2476–2486

Gnansounou E (2008) Fuel ethanol. Current status and outlook. In: Pandey A (ed) Handbook of plant-based biofuels. CRC Press, Boca Raton, pp 57–71

Ghose TK, Bisaria VS (1979) Studies on mechanism of enzymatic hydrolysis of cellulosic substances. Biotechnol Bioeng 21:131–146

Girio FM, Fonseca C, Carvalheiro F, Duarte CL, Marques S, Bogel-qukasik R (2010) Hemicelluloses for fuel ethanol: a review. Bioresour Technol 101:4775–4800

Godia F, Casas C, Sola C (1987) A survey of continuous ethanol fermentation systems using immobilized cells. Process Biochem 22–22:43–48

Goldemberg J (2007) Ethanol for a sustainable energy future. Science 315:808–810

Graf A and Koehler T (2000) Oregon cellulose-ethanol study. Oregon office of energy, Salem, OR USA, p 30 + appendices

Gray KA, Zhao L, Emptage M (2006) Bioethanol. Curr Opin Chem Biol 10:141–146

Guan W, Shi S, Tu M, Lee YY (2016) Acetone–butanol–ethanol production from Kraft paper mill sludge by simultaneous saccharification and fermentation. Bioresour Technol 2016(200):713–721

Gurram RN, Al-Shannag M, Lecher NJ, Duncan SM, Singsaas EL, Alkasrawi M (2015) Bioconversion of paper mill sludge to bioethanol in the presence of accelerants or hydrogen peroxide pretreatment. Bioresour Technol 192:529–539

Guzzon A, Bohn A, Diociaiuti M, Albertano P (2008) Cultured phototrophic biofilms for phosphorus removal in wastewater treatment. Water Res 42:4357–4367

Harner NK, Bajwa PK, Formusa PA, Austin GD, Habash MB, Trevors JT, Chan CK, Ho CY, Lee H (2015) Determinants of tolerance to inhibitors in hardwood spent sulfite liquor in genome shuffled Pachysolen tannophilus strains. Antonie Van Leeuwenhoek 108:811–834

Henriques T, Pereira S, Serafim L, Xavier A (2018) Two-stage aeration fermentation strategy to improve bioethanol production by Scheffersomyces stipitis. Fermentation 4:97

Ho NWY, Chen Z, Brainard A, Sedlak M (1999) Successful design and development of genetically engineered *Saccharomyces* yeasts for effective co fermentation of glucose and xylose from cellulosic biomass to fuel ethanol. Adv Biochem Eng Biotechnol 65:164–192

Ho NWY, Chen Z, Brainard AP (1998) Genetically engineered *Saccharomyces* yeast capable of effective co-fermentation of glucose and xylose. Appl Environ Microbiol 64:1852–1859

Hsu TA, Ladisch MR and Tsao GT (1980) Alcohol from cellulose. Chem Technol 10(5):315–319. https://doi.org/10.1016/j.biortech.2009.10.079

Hu Q, Kurano N, Kawachi M, Iwasaki I, Miyachi A (1998) Ultrahigh-cell-density culture of amarine alga Chlorococcum littorale in a flat-plate photobioreactor. Appl Microbiol Biotechnol 46:655–662

Huang R, Su R, Qi W, He Z (2011) Bioconversion of lignocellulose into bioethanol: process intensification and mechanism research. Bioenergy Res 4(4):225–245

İçöz E, Tuğrul KM, Saral A, İçöz E (2009) Research on ethanol production and use from sugar beet in Turkey. Biomass Bioenergy 33(1):1–7. https://doi.org/10.1016/j.biombioe.2008.05.005

Islam MA, Maitra PMd, Fakruddin Md, Abdul Q, Monzur MA, Naiyyum C (2012) Biotechnology 11(4):248–252

Jain S, Chouhan S, Chavan KJ, Singh P, Tiwari A (2016) Bioethanol production from waste paper sludge as a fermentation substrate by using xylose fermenting yeast. Minerva Biotecnol 28:7

Jeffries TW, Jin YS (2000) Ethanol and thermotolerance in the bioconversion of xylose by yeasts. Adv Appl Microbiol 47:221–268

Jiang L, Luo S, Fan X, Yang Z, Guo R (2011) Biomass and lipid production of marine microalgae using municipal waste water and high concentration of CO2. Appl Energy 88:3336–3341

Jobling S (2004) Improving starch for food and industrial applications. Curr Opin Plant Biol 7(2):210–218. https://doi.org/10.1016/j.pbi.2003.12.001

John RP, Anisha GS, Nampoothiri KM, Pandey A (2011) Micro and macroalgal biomass: a renewable source for bioethanol. Bioresour Technol 102(1):186–193

Johnson T, Johnson B, Scott-Kerr C and Kiviaho J (2010) Bioethanol—status report on bioethanol production from wood and other lignocellulosic feedstocks. In: 63rd Appita annual conference and exhibition, Melbourne, 19–22 Apr 2009

Johnson E (2016) Integrated enzyme production lowers the cost of cellulosic ethanol. Biofuels Bioprod Bioref 2016(10):164–174

Jones S, Zhu Y, Anderson D (2014) Process design and economics for the conversion of algal biomass to hydrocarbons: whole algae hydrothermal liquefaction and upgrading. Richland, WA: PNNL-23227

Jorquera O, Kiperstok A, Sales EA, Embiruçu M, Ghirardi ML (2010) Comparative energy life-cycle analyses of microalgal biomass production in open ponds and photobioreactors. Bioresour Technol 101(4):1406–1413

Khanal SK (2008) Anaerobic biotechnology for bioenergy production: principles and applications. Wiley-Blackwell, Ames, IA, USA. https://doi.org/10.1002/9780813804545

Kim S, Dale BE (2004a) Cumulative energy and global warming impact from the production of biomass for biobased products. J Indust Ecol 7(3–4):147–162

Kim S, Dale BE (2004b) Global potential bioethanol production from wasted crops and crop residues. Biomass Bioenergy 26:361–375

Kim G (2018) China Biofuels Annual 2018; Global Agricultural Information Network (GAIN) Report Number: CH18041; US Department of Agriculture (USDA) Foreign Agricultural Service: Washington. DC, USA

Klinke HB, Thomsen AB, Ahring BK (2004) Inhibition of ethanol producing yeast and bacteria by degradation products produced during pretreatment of biomass. Appl Microbiol Biotechnol 66:10–26

Ko CH, Wang YN, Chang FC, Chen JJ, Chen WH, Hwang WS (2012) Potentials of lignocellulosic bioethanols produced from hardwood in Taiwan. Energy 2012(44):329–334

Krishna SH, Reddy TJ, Chowdary GV (2001) Simultaneous saccharification and fermentation of lignocellulosic wastes to ethanol using thermotolerant yeast. Bioresour Technol 77:193–196

Kroumov AD, M′odenes AN and de Araujo Tait MC (2006) Development of new unstructured model for simultaneous saccharification and fermentation of starch to ethanol by recombinant strain. Biochem Eng J 28:243–255

Kumar R, Tabatabaei M, Karimi K, Sárvári Horváth I (2016) Recent updates on lignocellulosic biomass derived ethanol—a review. Biofuel Res J 9:347–356

Ladisch MR (2003) Apollo program for biomass liquids what will it take? Universiteit Utrecht Copernicus Institute, Science Technology Society. www.purdue.edu/energysummit/presentations/ladisch_purdue.pdf

Lane J (2017) New life for INEOS Bio plant: Alliance Bio-Products Wins US OK for Cellulosic Ethanol Re-fit. Biofuels Digest. http://www.biofuelsdigest.com/bdigest/2017/07/11/new-life-for-ine

Lane S (2016) Sinatra Bio: Ol' Brew Eyes Is back. Biofuels Digest. http://www.biofuelsdigest.com/bdigest/tag/synata-bio/

Larsson S, Palmqvist E, Hahn Hagerdal B, Tengborg C, Stenberg K, Zacchi G, Nilvevrant NO (1999) The generation of fermentation inhibitors during dilute acid hydrolysis of soft wood. Enzyme MicrobTechnol 24:151–159

Lawford HG, Rousseau JD (1993). Production of ethanol from pulp mill hardwood and softwood spent sulfite liquors by genetically engineered E. coli. Appl Biochem Biotechnol 39–40, 667–685

Lee S, Speight JG, Loyalka SK (2007) Handbook of alternative fuel technologies. CRC Press, Boca Raton, FL, USA. DOI: https://doi.org/10.1201/

Lehr F, Posten C (2009) Closed photo-bioreactors as tools for biofuel production. Curr Opin Biotechnol 20(3):280–285

Lennartsson PR, Erlandsson P, Taherzadeh MJ (2014) Integration of the first and second generation bioethanol processes and the importance of byproducts. Bioresour Technol 165:3–8

Li K, Liu S, Liu X (2014) An overview of algae bioethanol production. Int J Energy Res 38(8):965–977. https://doi.org/10.1002/er.3164

Linden T, Hahn-Higerdal B (1989) Fermentation of lignocellulose hydrolysates with yeasts and xylose isomerase. Enzyme Microb Technol 11:337–353

Ling KC (2008) Whey to ethanol: A biofuel role for dairy cooperatives? USDA Rural Development: Research Report 214. http://www.rd.usda.gov/files/RR214.pdf

Linoj KNV, Dhavala P, Goswami A, Maithel S (2006) Liquid biofuels in South Asia: resources and technologies. Asian Biotechnol Dev Rev 8(2):31–49

Liu CG, Xiao Y, Xia XX, Zhao XQ, Peng L, Srinophakun P, Bai FW (2019) Cellulosic ethanol production: progress, challenges and strategies for solutions. Biotechnol Adv 37:491–504

Lopez Bautista J, Rindi MF, Guiry MD (2006) Molecular systematics of the sub aerial green algal order Trentepohliales: an assessment based on morphological and molecular data. Int J Syst Evol Microbiol 2006(56):1709–1715

Lopez MJ, Nichols NN, Dien BS, Moreno J, Bothast RJ (2004) Isolation of microorganisms for bio logical detoxification of lignocellulosic hydrolysates. Appl Microb Biotechnol 64:125–131

Lorenz D and Morris D (1995) How much energy does it take to make a gallon of ethanol? Institute for Local Self-Reliance, Aug 1995

Lundquist TJ, Woertz IC, Quinn NW, Benemann JR (2010) A realistic technology and engineering assessment of algae biofuel production. Energy Biosciences Institute, University of California, Berkeley, CA

Lynd LR (1996) Overview and evaluation of fuel ethanol from cellulosic biomass: technology, 0economics, the environment, and policy. Annu Rev Energy Environ 21:403–465

Lynd LR, van Zyl WH, McBride JE, Laser M (2005) Consolidated bioprocessing of cellulosic biomass: an update. Curr Opin Biotechnol 16:577–583

Lynd LR, Wyman CE, Gerngross TU (1999) Biocommodity engineering. Dartmouth College/Thayer School of Engineering, Hanover

Mallick N, Mandal S, Singh AK, Bishai M, Dash A (2012) Green microalga Chlorella vulgaris as a potential feedstock for biodiesel. J Chem Technol Biotechnol 87(1):137–145

Mandal S, Mallick N (2009) Microalga Scenedesmus obliquus as a potential source for biodiesel production. Appl Microbiol Biotechnol 84:281–291

Mandenius CF (1988) Controlling fermentation of lignocellulose hydrolysates in a continuous hollow-fiber reactor using biosensors. Biotechnol Bioeng 32:123–129

Markou G, Angelidaki I, Nerantzis E, Georgakakis D (2013) Bioethanol production by carbohydrate-enriched biomass of Arthrospira (Spirulina) platensis. Energies 2013(6):3937–3950

Marko-Varga G, Buttler T, Gorton L, Olsson L, Durand G, Barceló D (1994) Qualitative and quantitative carbohydrate analysis of fermentation substrates and broths by liquid chromatographic techniques. J Chromatogr A 665:317–332

Marques S, Alves L, Roseiro JC, Gírio FM (2008) Conversion of recycled paper sludge to ethanol by SHF and SSF using Pichia stipitis. Biomass Bioenergy 32:400–406

Martín C, Galbe M, Wahlbom CF, Hägerdal BH, Jönsson LJ (2002) Ethanol production from enzymatic hydrolysates of sugarcane bagasse using recombinant xylose-utilising *Saccharomyces cerevisiae*. Enz Microb Technol 31:274–282

Martinez A, Rodriguez ME, York SW, Preston JF, Ingram LO (2000) Effects of Ca(OH)2 treatments ("overliming") on the composition and toxicity of bagasse hemicellulose hydrolysates. Biotechnol Bioeng 69:526–536

Matthew H, Ashley O, Brian K, Alisa E, Benjamin JS (2005) Wine making 101

McGlashen A (2013) As key partner departs, future dims for Michigan cellulosic biofuel plant. https://energynews.us/2013/08/06/midwest/as-key-partnerdeparts-future-dims-for-michigancellulosic-biofuel-plant/

Mendes CVT, Cruz CHG, Reis DFN, Carvalho MGVS, Rocha JMS (2016) Integrated bioconversion of pulp and paper primary sludge to second generation bioethanol using Saccharomyces cerevisiae ATCC 26602. Bioresour Technol 220:161–167

Mendes CVT, Rocha JM, dos S, de Menezes FF, Carvalho M, da GVS (2017) Batch and fed-batch simultaneous saccharification and fermentation of primary sludge frompulp and paper mills. Environ Technol 38:1498–1506

Mendes CVT, Rocha JMS, Carvalho MGVS (2014) Valorization of residual streams from pulp and paper mills: Pretreatment and bioconversion of primary sludge to bioethanol. Ind Eng Chem Res 53:19398–19404

Mendez L, Mahdy A, Ballesteros M, Gonzalez-Fernandez C (2014) Methane production of thermally pretreated *Chlorella vulgaris* and *Scenedesmus* sp. biomass at increasing biomass loads. Appl Energy 129:238–242

Moazami N, Ashori A, Ranjbar R, Tangestani M, Eghtesadi R, Nejad AS (2012) Largescale biodiesel production using microalgae biomass of Nannochloropsis. Biomass-Bioenergy 39(2012):449–453

Modahl IS, Brekke A, Valente C (2015) Environmental assessment of chemical products from a Norwegian biorefinery. J Clean Prod 94:247–259

Mohandas DV, Whelan DR, Panchal CJ (1995) Development of xylose-fermenting yeasts for ethanol production at high acetic acid concentrations. Appl Biochem Biotechnol 51–52:307–318

Molina GE, Belarbi EH, Fernandez FG, Medina RA, Chisti Y (2001) Tubular photobioreactor design for algal cultures. J Biotechnol 92(2):113–131

Monrroy M, García JR, Mendonça RT, Baeza J, Freer J (2012) Kraft pulping of Eucalyptus globulus as a pretreatment for bioethanol production by simultaneous saccharification and fermentation. J Chil Chem Soc 57:1113–1117

Morris D (1995) How much energy does it take to make a gallon of ethanol? Institute for Local Self-Reliance. www.carbohydrateeconomy.org/library/admin/uploadedfiles/How_Much_Energy_Does_it_Take_to_Make_a_Gallon_html-33k

Mosier N, Wyman C, Dale B, Elander R, Lee YY, Holtzapple M, Ladisch M (2005) Features of promising technologies for pretreatment of lignocellulosic biomass. Bioresour Technol 96:673–686

Mousdale DM (2008). Biofuels. Biotechnology, chemistry and sustainable development. CRC Press, Boca Raton

Mussatto SI1, Dragone G, Guimarães PM, Silva JP, Carneiro LM, Roberto IC, Vicente A, Domingues L, Teixeira JA (2010) Technological trends, global market, and challenges of bioethanol production. Biotechnol Adv 28(6):817–830. https://doi.org/10.1016/j.biotechadv.2010.07.001

Nagarajan S, Chou SK, Cao S, Wu C, Zhou Z (2013) An updated comprehensive technoeconomic analysis of algae biodiesel. Bioresour Technol 2013(45):150–156
Nair SU, Ramachandran S, Pandey A (2008) Bioethanol from starch biomass. Part 1. Production of starch saccharifying enzymes. In: Pandey A (ed) Handbook of plant-based biofuels. CRC Press, Boca Raton, pp 87–103
Najafpour GD (1990) Immobilization of microbial cells for the production of organic acids. J Sci Islam Repub Iran 1:172–176
Neste Oil Corporation Neste (2013) Oil signs an algae oil off-take agreement with Cellana. https://www.neste.com/en/neste-oil-signs-algae-oil-takeagreement-cellana
Neste Oil Corporation Neste (2014) Oil strengthens its algae oil procurement program with a new off-take agreement. https://www.neste.com/en/nesteoil-strengthens-its-algae-oil-procurement-program-new-take-agreement
Nguyen Q, Bowyer J, Howe J, Bratkovich S, Groot H, Pepke E, Fernholz K (2017) Global production of second generation biofuels: trends and influences. https://dovetailinc.org/upload/tmp/1579558792.pdf
Nigam JN (2002) Bioconversion of water-hyacinth (*Eichhornia crassipes*) hemicellulose acid hydrolysate to motor fuel ethanol by xylose–fermenting yeast. J Biotechnol 97:107–116
Nigam JN (2001) Ethanol production from hardwood spent sulfite liquor using an adapted strain of Pichia stipitis. J Ind Microbiol Biotechnol 2001(26):145–150
Nilvebrant N, Reimann A, Larsson S, Jonsson LJ (2001) Detoxification of lignocellulose hydrolysates with ion exchange resins. Appl Biochem Biotechnol 91–93:35–49
Niphadkar S, Bagade P, Ahmed S (2017) Bioethanol production: insight into past, present and future perspectives. Biofuels. https://doi.org/10.1080/17597269.2017.1334338
Norsker NH, Barbosa MJ, Vermuë MH, Wijffels RH (2011) Microalgal production—a close look at the economics. Biotechnol Adv 2011(29):24–27
O'Brien D, Woolverton M (2009) Recent trends in U.S. wet and dry corn milling production. AgMRC Renewable Newsletter. http://www.agmrc.org/renewable_energy
Olsson L, Hahn-Hägerdal B (1996) Fermentation of lignocellulosic hydrolysates for ethanol production. Enzyme Microb Technol 18:312–331
Olsson L, Jorgensen H, Krogh KBR and Roca C (2005) Bioethanol production from lignocellulosic material. In: Dumitriu S (ed) Polysaccharides. Structural diversity and functional versatility, 2nd edn, Chap 42. Marcel Dekker, New York, pp 957–993
Olsson L, Hahn-Hägerdal B (1993) Fermentative performance of bacteria and yeasts in lignocellulose hydrolysates. Process Biochem 28:249–257
Padella M, Edwards R, Candela Ripoll I, Lonza L (2018) Estimates of Biofuel Production Costs and Cost of Savings; JEC Well-to-Tank Version 5, Annex 5; European Commission: Ispra, Italy; JRC115003
Padella M, O'Connell A, Prussi M (2019) What is still limiting the deployment of cellulosic ethanol? Analysis of the current status of the sector. Appl Sci 9:4523
Palmqvist E, Hahn-Hagerdal B (2000) Fermentation of lignocellulosic hydrolysates. I: inhibition and detoxification and II: inhibitors and mechanisms of inhibition. Bioresour Technol 74:17–33
Palmqvist E, Hahn-Hägerdal B, Galbe M, Zacchi G (1996) The effect of water-soluble inhibitors from steam-pretreated willow on enzymatic hydrolysis and ethanol fermentation. Enzyme Microb Technol 19:470–476
Palmqvist E, Galbe M, Hahn-Hägerdal B (1998) Evaluation of cell recycling in continuous fermentation of enzymatic hydrolysates of spruce with Saccharomyces cerevisiae and on-line monitoring of glucose and ethanol. Appl Microbiol Biotechnol 50:545–551
Pandey A, Nigam P, Soccol CR, Soccol VT, Singh D, Mohan R (2000) Advances in microbial amylases. Biotechnol Appl Biochem 31(Pt 2):135–152. https://doi.org/10.1042/BA19990073
Patil VKQ, Giselrod HR (2008) Towards sustainable production of biofuels from microalgae. Biol Environ 9:1188–1195

Pavlečić M, Rezić T, Ivančić Šantek M, Horvat P, Šantek B (2017) Bioethanol production from raw sugar beet cossettes in horizontal rotating tubular bioreactor. Bioprocess Biosyst Eng 40(11):1679–1688. https://doi.org/10.1007/s00449-017-1823-x

Pavlečić M, Vrana I, Vibovec K, Ivančić Šantek M, Horvat P, Šantek B (2010) Ethanol production from different intermediates of sugar beet processing. Food Technol Biotechnol. 48(3):362–367

Pelli P, Haapala A, Pykäläinen J (2017) Services in the forest-based bioeconomy—analysis of European strategies. Scand J For Res 32:559–567

Peng L, Chen Y (2011) Conversion of paper sludge to ethanol by separate hydrolysis and fermentation (SHF) using Saccharomyces cerevisiae. Biomass Bioenergy 35:1600–1606

Pereira SR, Ivanuša Š, Evtuguin DV, Serafim LS, Xavier AMRB (2012) Biological treatment of eucalypt spent sulphite liquors: a way to boost the production of second generation bioethanol. Bioresour Technol 2012(103):131–135

Pereira SR, Portugal-Nunes DJ, Evtuguin DV, Serafim LS, Xavier AMRB (2013) Advances in ethanol production from hardwood spent sulphite liquors. Process Biochem 48:272–282

Pereira SR, Sànchez i Nogué V, Frazão CJR, Serafim LS, Gorwa-Grauslund MF, Xavier AMRB (2015) Adaptation of Scheffersomyces stipitis to hardwood spent sulfite liquor by evolutionary engineering. Biotechnol Biofuels 8

Petrou EC, Pappis CP (2009) Biofuels: a survey on pros and cons. Energy Fuels 23:1055–1066

Phillips RB, Jameel H, Chang HM (2013) Integration of pulp and paper technology with bioethanol production. Biotechnol Biofuels 6:13

Pimental D (2002) Limits of biomass utilization. Encyclopedia of physical science and technology. 3rd edn, vol 2. Academic Press, pp 159–171

Pinel D, D'Aoust F, del Cardayre SB, Bajwa PK, Lee H, Martin VJJ (2011) Saccharomyces cerevisiae genome shuffling through recursive population mating leads to improved tolerance to spent sulfite liquor. Appl Environ Microbiol 2011(77):4736–4743

Pirt SJ, Lee YK, Walach MR, Pirt MW, Balyuzi HHM, Bazin MJ (1983) A tubular bioreactor for photosynthetic production of biomass from carbon dioxide: design and performance. J Chem Technol Biotechnol 1983(33B):35–58

Piskur J, Rozpedowska E, Polakova S, Merico A, Compagno C (2006) How did *Saccharomyces* evolve to become a good brewer? Trends Genet 22(4):183–186

POET-DSM: Project Liberty (2018) https://www.energy.gov/eere/bioenergy/poet-dsm-project-liberty, https://www.forbes.com/sites/rrapier/2018/02/11/cellulosicethanol-falling-far-short-ofthehype/#484c7df4505f

Portugal-Nunes D, Sànchez i Nogué V, Pereira, SR, Craveiro, SC, Calado AJ, Xavier AM (2015) Effect of cell immobilization and pH on Scheffersomyces stipitis growth and fermentation capacity in rich and inhibitory media. Bioresour. Bioprocess 2:13. https://doi.org/10.1186/s40643-015-0042-z

Prasad S, Singh A. Jain N, Joshi HC (2007) Ethanol production from hum syrup for utilization as automotive fuel in India. Energy Fuel 21(4):2415–2420

Pulz O (2001) Photobioreactors: production systems for phototrophic microorganisms. Appl Microbiol Biotechnol 57:287–293

Queirós D, Fonseca A, Lemos PC, Serafim LS (2016) Long-term operation of a two-stage polyhydroxyalkanoates production process from hardwood sulphite spent liquor. J Chem Technol Biotechnol 2016(91):2480–2487

Queirós D, Rossetti S, Serafim LS (2014) PHA production by mixed cultures: a way to valorize wastes from pulp industry. Bioresour Technol 2014(157):197–205

Queirós D, Sousa R, Pereira S, Serafim LS (2017) Valorization of a pulp industry by-product through the production of short-chain organic acids. Fermentation 2017(3):20

Quintero JA, Rincón LE, Cardona CA (2011) Production of bioethanol from agroindustrial residues as feedstocks. In: Pandey A, Larroche C, Ricke SC, Dussap CG, Gnansounou E, eds, Biofuels—Alternative feedstocks and conversion processes. Academic Press, Oxford, UK, pp. 251–285

Raes EJ, Isdepsky A, Muylaert K, Borowitzka MA, Moheimani NR (2014) Comparison of growth of Tetraselmis in a tubular photobioreactor (Biocoil) and a raceway pond. J Appl Phycol 2014(26):247–255

Rajkumar R, Yaakob Z, Takriff MS (2014) Potential of micro algae and macro algae for biofuel production: a brief review. BioResources 9(1):1606–1633

Ren HY, Liu BF, Ma C, Zhao L, Ren NQ (2013) A new lipid rich microalga Scenedesmus sp. strain R-16 isolated using Nile red staining: effects of carbon and nitrogen sources and initial pH on the biomass and lipid production. Biotechnol Biofuels 6:143–152

RFA-Renewable Fuels Association (2018) Ethanol strong, Ethanol industry outlook. https://ethanolrfa.org/wp-content/uploads/2018/02/NECfinalOutlook.pdf

RFA—Renewable Fuels Association (2007a) Ethanol industry outlook: building new horizons. http://www.ethanolrfa.org/objects/pdf/outlook/RFA_Outlook_2007.pdf

RFA—Renewable Fuels Association (2007b) Cellulosic ethanol grants provide much needed boost to fledgling technology. www.biofuelsjournal.com/articles/RFA

Richardson JW, Johnson MD, Lacey R, Oyler J, Capareda S (2014) Harvesting and extraction technology contributions to algae fuels economic viability. Algal Res 5:70–78

Richmond A, ChengWZ and Zarmi Y (2003). Efficient use of strong light for high photosynthetic productivity: interrelationships between the optical path, the optimal population density and cell-growth inhibition. Biomol Eng 20(4–6):229–236

Ridout A, Boshoff S, Haigh K, Carrier M, Van E, Görgens J (2016) Valorisation of paper waste sludge via fermentation and pyrolysis. Cellul Chem Technol 2016(50):439–442

Roadmap B (2012) German Federal Government action plans for the material and energetic utilisation of renewable raw materials. https://www.bmbf.de/pub/Roadmap_Biorefineries_eng.pdf

Rodolfi L, Zittelli GC, Bassi N, Padovani GN, Biondi G, Bonini G, Tredici MR (2009) Microalgae for oil: strain selection, induction of lipid synthesis and outdoor mass cultivation in a low-cost photobioreactor. Biotechnol Bioeng 102:100–112

Rødsrud G, Lersch M, Sjöde A (2012) History and future of world's most advanced biorefinery in operation. Biomass Bioenergy 46:46–59

Rogers JN, Rosenberg JN, Guzman BJ (2014) A critical analysis of paddlewheel-driven raceway ponds for algal biofuel production at commercial scales. Algal Res 2014(4):76–88

Rueda C, Calvo PA, Moncalián G, Ruiz G, Coz A (2015) Biorefinery options to valorize the spent liquor from sulfite pulping. J Chem Technol Biotechnol 90:2218–2226

Rueda C, Marinova M, Paris J, Ruiz G, Coz A (2016) Technoeconomic assessment of different biorefinery approaches for a spent sulfite liquor. J Chem Technol Biotechnol 2016(91):2646–2653

Safi BF, Rouleau D, Mayer RC, Desrochers M (1986) Fermentation kinetics of spent sulfite liquor by Saccharomyces cerevisiae. Biotechnol Bioeng 28:944–951

Saha BC, Iten LB, Cotta MA, Wu YV (2005) Dilute acid pretreatment, enzymatic saccharification and fermentation of wheat straw to ethanol. Proc Biochem 40:3693–3700

Sapp M (2018) Eni's versalis wins biochemtex and beta renewables at auction. Biofuels Digest. http://www.biofuelsdigest.com/bdigest/2018/10/01/enis-versalis-wins-biochemtex-and-beta-renewables-at-auction/

Sassner P, Galbe M, Zacchi G (2008) Techno-economic evaluation of bioethanol production from three different lignocellulosic materials. Biomass Bioenergy 32:422–430

Schill RS (2018) Zero to 10 Million in 5 Years. Ethanol Producers Magazine. http://www.ethanolproducer.com/articles/15344/zero-to-10-million-in-5-years

Schmidt SK, Lynch RC, King AJ, Karki D, Robeson MS, Nagy L, Williams MW, Mitter MS, Freeman KR (2011) Phylogeography of microbial phototrophs in the dry valleys of the high Himalayas and Antarctica. Proc R Soc B: Biol Sci 2011(278):702–708

Schneider HI (1989) Conversion of pentoses to ethanol by yeast and fungi. Crit Rev Biotechnol 9:1–40

Schneider H (1996) Selective removal of acetic acid from hardwood-spent sulfite liquor using a mutant yeast. Enzyme Microb Technol 1996(19):94–98

Schroeder BG, Zanoni PRS, Magalhães WLE, Hansel FA, Tavares LBB (2017) Evaluation of biotechnological processes to obtain ethanol from recycled paper sludge. J Mater Cycles Waste Manag 2017(19):463–472

Schubert C (2006) Can biofuels finally take center stage? Nat Biotechnol 24:777–784

Searle S, Pavlenko N, El Takriti S and Bitnere K (2017) Potential Greenhouse Gas Savings from a 2030 Greenhouse GasReduction Target with Indirect Emissions Accounting for the European Union;Working Paper 2017-05; International Council on Clean Transportation (ICCT): Washington, DC, USA

Senthilkumar V, Gunasekaran P (2008) Bioethanol from biomass. Production of ethanol from molasses. In: Pandey A (ed) Handbook of plant-based biofuels. CRC Press, Boca Raton, pp 73–86

Shapouri H, Duffield JA and Graboski MS (1995) Estimating the net energy balance of corn ethanol. AER-721. USDA Economic Research Service, Washington, DC

Shapouri H, Duffield JA and Wang M (2002) The energy balance of corn ethanol: an update. AER-814. USDA Office of the Chief Economist, Washington, DC

Shapouri H, Duffield JA, Wang M (2003) The energy balance of corn ethanol revisited. Am Soc Agric Eng 46(4):959–968

Shapouri H, McAloon A (2004) The 2001 net energy balance of corn ethanol. U.S Department of Agriculture, Washington, DC

Shigechi H, Koh J, Fujita Y, Matsumoto T, Bito Y, Ueda M, Satoh E, Fukuda H, Kondo A (2004) Direct production of ethanol from raw corn starch via fermentation by use of a novel surface-engineered yeast strain codisplaying glucoamylase and α-amylase. Appl Environ Microbiol 70(8):5037–5040. https://doi.org/10.1128/AEM.70.8.5037-5040.2004

Shilton AN, Powell N, Mara DD, Craggs R (2008) Solar-powered aeration and disinfection, anaerobic co-digestion, biological CO2 scrubbing and biofuel production: the energy and carbon management opportunities of waste stabilization ponds. Water Sci Technol 58:253–258

Sileveria WB, Passos FJV, Mantovani HC, Paasos FML (2005) Ethanol production from cheese whey permeate by Klyveromyces marxianus UFV-3: A flux analysis of oxido-reductive metabolisms as a function of lactose cancentration and oxygen levels. Enzyme Microb Technol 36:930–936. https://doi.org/10.1016/j.enzmictec.2005.01.018

Sivers MV, Zacchi G (1995) A techno-economical comparison of three processes for the production of ethanol from pine. Bioresour Technol 51:43–52

Singh A, Kumar PK (1991) *Fusarium oxysporum*: status in bioethanol production. Crit Rev Biotechnol 11(2):129–147

Singh A, Pant D, Korres NE, Nizami AS, Prasad PS and Murphy JD (2010). Key issues in life cycle assessment of ethanol production from lignocellulosic biomass: challenges and perspectives. Bioresour Technol 2010;101(13):5003–5012

Sjöström, E (1993). Wood Chemistry: Fundamentals and Applications, 2nd ed, Academic Press: San Diego, CA, USA,; pp. 137, 138, 158, 238, 240, 244, ISBN 978-0-12-647481-7

Soccol CR, Faraco V, Karp S, Vandenberghe LPS, Thomaz-Soccol V, Woiciechowski A, and Pandey A (2011). Lignocellulosic bioethanol: Current status and future perspectives. In: Pandey A, Larroche C, Ricke SC, Dussap CG, Gnansounou E, editors. Biofuels – Alternative feedstocks and conversion processes. Oxford, UK: Academic Press; 2011. pp. 101–122. https://doi.org/10.1016/B978-0-12-385099-7.00005-X

Solomon BD, Barnes JR, Halvorsen KE (2007) Grain and cellulosic ethanol: History, economics, and energy policy. Biomass Bioenergy 31(6):416–425. https://doi.org/10.1016/j.biombioe.2007.01.023

Sree NK, Sridhar M, Suresh K, Banat IM, Rao LV (2000) Isolation of thermotolerant, osmotolerant, flocculating *Saccharomyces cerevisiae* for ethanol production. Biores Technol 72:43–46

Sun Y, Cheng J (2002) Hydrolysis of lignocellulosic materials for ethanol production: a review. Bioresour Technol 83:1–11

Tabak J (2009) Biofuels. Infobase Publishing, New York, NY, USA

Taherjadeh M (1999) Ethanol from lignocellulose: physiological effects of inhibitors and fermentation strategies. PhD thesis. Lund University, Lund, Sweden

Takagi M, Karseno and Yoshida T (2006). Effect of salt concentration on intracellular accumulation of lipids and triacylglyceride in marine microalgae Dunaliella cells. J Biosci Bioenergy 101:223–226

Tanadul OU, VanderGheynst JS, Beckles DM, Powell ALT, Labavitch JM (2014) The impact of elevated CO2 concentration on the quality of algal starch as a potential biofuel feedstock. Biotechnol Bioeng 111(7):1323–1331

Tanaka L (2006) Ethanol fermentation from biomass resources: current state and prospects. Appl Microbiol Biotechnol 69:627–642

Thamsiriroj T, Murphy JD (2009) A technical analysis of bioenergy production from oilseed rape. In: Proceedings of the 3rd international conference on sustainable energy and environment (SEE). World Renewable Energy Congress 2009-Asia, Bangkok, Thailand

Tian S, Zhou G, Yan F, Yu Y, Yang X (2009) Yeast strains for ethanol production from lignocellulosic hydrolysates during in situ detoxification. Biotechnol Adv 27:656–660

Tran TTA, Le TKP, Mai TP, Nguyen DQ (2019) Bioethanol production from lignocellulosic biomass. In: Alcohol fuels-current technologies and future prospect. IntechOpen, London

US Department of Energy (2010) Office of energy efficiency and renewable energy, Biomass Program

US DOE (2003) National algal biofuels technology roadmap. https://www1.eere.energy.gov/bioenergy/pdfs/algal_biofuels_roadmap.pdf

Ugwu CU, Aoyagi H, Uchiyama H (2008) Photobioreactors for mass cultivation of algae. Bioresour Technol 99:4021–4028

Valin H, Peters D, van den Berg M, Frank S, Havlik P, Forsell N, Hamelinck C (2015) The land use change impact of biofuels consumed in the EU. Quantification of area and greenhouse gas impacts, Ecofys, Utrecht, the Netherlands

Vandamme D, Pontes SCV, Goiris K, Foubert I, Pinoy LJ, Muylaert K (2011) Evaluation of electro-coagulation-flocculation for harvesting marine and freshwater microalgae. Biotechnol Bioeng 108:2320–2329

Verbio Press Release (2018) VERBIO to Acquire DuPont's Nevada, Iowa-Based Cellulosic Ethanol Plant. Press-release. https://www.verbio.de/en/press/news/press-releases/verbio-to-acquire-duponts-nevada-iowa-based-cellulosic-ethanol-plant/

Vinogradova OM, Darienko TM (2008) Terrestrial algae of hypersaline environments of the Central Syvash islands (Kherson Region, Ukraine). Biologia 63:813–823

Voegele E (2014) Bluefire announces EPC contract for mississippi project (2014). http://ethanolproducer.com/articles/11521/bluefire-announcesepc-contract-for-mississippi-project

Voegele E (2018). Verbio to buy DuPont cellulosic ethanol plant, convert it to RNG. 2018. http://biomassmagazine.com/articles/15743/verbio-to-buy-dupontcellulosic-ethanolplant-convert-it-to-rng

Voorhis D (2016) Hugoton cellulosic ethanol plant sold out of bankruptcy. 2016. https://www.kansas.com/news/business/article119902263.html

Walker GM (2010) Bioethanol: Science and technology of fuel alcohol. Ventus Publishing ApS ISBN 978-87-7681-681-0

Wang M (1999) Argonne National Laboratory, biofuels: energy balance. http://www.iowacorn.org/ethanol/documents/energy_balance_000.pdf

Wang M (2005) An update of energy and greenhouse emission impacts of fuel ethanol. Center for Transportation Research Argonne National Laboratory, The 10th Annual National Ethanol Conference Scottsdale, AZ. http://www.ethanol-gec.org/netenergy/UpdateEnergyGreenhouse.pdf

Wang M, Saricks C, Santini D (1999) Effects of fuel ethanol on fuel-cycle energy and greenhouse gas emissions. Argonne National Laboratory. ANL/ESD-38, p 39. http://www.transportation.anl.gov/pdfs/TA/58.pdf

Wang Y, Guo W, Chen BY, Cheng CL, Lo YC, Ho SH, Chang JS, Ren N (2015) Exploring the inhibitory characteristics of acid hydrolysates upon butanol fermentation: a toxicological assessment. Bioresour Technol 198:571–576

Watanabe Y and Saiki H (1997) Development of a photobioreactor incorporating Chlorella sp. for removal of CO2 in stack gas. Energy Convers Manag 38:499–503

Wiley PE, Brenneman KJ, Jacobson AE (2009) Improved algal harvesting using suspended air flotation. Water Environ Res 81:702–708

Wingren A, Galbe M, Zacchi G (2003) Techno-economic evaluation of producing ethanol from softwood: comparison of SSF and SHF and identification of bottlenecks. Biotechnol Prog 19:1109–1117

Wistara NJ, Pelawi R, Fatriasari W (2016) The effect of lignin content and freeness of pulp on the bioethanol productivity of Jabon wood. Waste Biomass Valorization 7:1141–1146

Wooley R, Ruth M, Sheehan J, Ibsen K, Majdeski H, Galvez A (1999) Lignocellulosic biomass to ethanol—process design and economics utilizing co-current dilute acid prehydrolysis and enzymatic hyrolysis—current and futuristic scenarios. National Renewable Energy Laboratory, Golden Colorade USA, p 72 + annexes

Wu S, Chang H, Phillips R, Jameel H (2014) Techno-economic analysis of the optimum softwood lignin content for the production of bioethanol in a repurposed Kraft mill. BioResources 9(4):6817–6830

Wuertz S, Bishop PL, Wilderer PA (2003) Biofilms in wastewater treatment: an interdisciplinary approach. IWA Publishing

Wyman CE, Bain RL, Hinman ND, Stevens DJ (1993) Ethanol and methanol from cellulosic biomass. In: Johansson TB, Kelly H, Reddy AKN, Williams RH, Burnham L (eds) Renewable energy, sources for fuels and electricity. Island Press, Washington DC, pp 865–923

Wyman CE, Dale BE, Elander RT, Holtzapple M, Ladisch MR, Lee YY (2005) Comparative sugar recovery data from laboratory scale application of leading pretreatment technologies to corn stover. Biores Technol 96(18):2026–2032

Xavier AMRB, Correia MF, Pereira SR, Evtuguin DV (2010) Second-generation bioethanol from eucalypt sulphite spent liquor. Bioresour Technol 101:2755–2761

Yamada T, Fatigati MA, Zhang M (2002) Performance of immobilized Zymomonas mobilis 31821 (pZB5) on actual hydrolysates produced by Arkenol technology. Appl Biochem Biotechnol 98:899–907

Yang IS, Salama ES, Kim JO, Govindwar SP, Kurade MB, Lee M, Roh HS, Jeon BH (2011a) Cultivation and harvesting of microalgae in photobioreactor for biodiesel production and simultaneous nutrient removal. Energy Convers Manag 2016(117):54–62

Yang J, Li X, Hu H, Zhang X, Yu Y, Chen Y (2011b) Growth and lipid accumulation properties of a freshwater microalga, Chlorella ellipsoidea in domestic secondary effluents. Appl Energi 88:3295–3299

Yang IS, Salama ES, Kim JO, Govindwar SP, Kurade MB, Lee M, Roh HS, Jeon BH (2016) Cultivation and harvesting of microalgae in photobioreactor for biodiesel production and simultaneous nutrient removal. Energy Conv Manag 117:54–62

Yu S, Wayman M, Parekh SK (1987) Fermentation to ethanol of pentose-containing spent sulphite liquor. Biotechnol Bioeng 29:1144–1150

Zafar S, Owais M (2006) Ethanol production from crude whey by Kluyveromyces marxianus. Biochem Eng J 2006(27):295–298

Zechman FW, Verbruggen H, Leliaert F, Ashworth M, Buchheim MA, Fawley MW, Spalding H, Pueschel CM, Buchheim JA, Verghese B, Hanisak MD (2010) An unrecognized ancient lineage of green plants persists in deep marine waters. J Phycol 46:1288–1295

Zettler LA, Gomez FE, Zettler E, Keenan B, Amils GR, Sogin ML (2002) Eukaryotic diversity in Spain's River of Fire. Nature 417:137

Zhang C (2019) Lignocellulosic Ethanol: technology and economics. In: Y Yun (ed) Alcohol fuels current technologies and future prospect. IntechOpen Limited, London. https://doi.org/10.5772/intechopen.86701

Zhou A, Thomson E (2009) The development of biofuels in Asia. Appl Energy 86:11–20
Zhou W, Chen P, Min M (2014) Environment-enhancing algal biofuel production using wastewaters. Renew Sustain Energy Rev 2014(36):256–269
Zhu LD, Hiltunen E, Antila E, Zhong JJ, Yuan ZH, Wang ZM (2014) Microalgal biofuels: flexible bioenergies for sustainable development. Renew Sustain Energy Rev 30:1035–1046

Link.springer.com

Aro EM (2016) From first generation biofuels to advanced solar biofuels. Ambio 45(Suppl. 1):S24–S31. https://doi.org/10.1007/s13280-015-0730-0
Bajpai P (2019) Biomass to energy conversion technologies-the road to commercialization, 1st edn. Elsevier
Berla BM, Saha R, Immethun CM, Maranas CD, Moon TS, Pakrasi HB (2013) Synthetic biology of cyanobacteria: unique challenges and opportunities. Front Microbiol 4:246. https://doi.org/10.3389/fmicb.2013.00246
Cameron DE, Bashor CJ, Collins JJ (2014) A brief history of synthetic biology. Nat Rev Microbiol 12:381–390
Christianson and Associates (2016) Ethanol evolution: the data and deas driving the future, J Christianson, principal partner, Christianson & Associates, PLLP, and Mark Fisler, Managing Director, Ocean Park Advisors. Presentation at Renewable Fuels Association National Ethanol Conference, New Orleans, LA, 17 Feb 2016
Gallagher P, Yee W, Baumes HS (2016) 2015 energy balance for the Corn-Ethanol industry, office of energy policy and new uses. Department of Agriculture, US
Hays SG, Ducat DC (2015) Engineering cyanobacteria as photosynthetic feedstock factories. Photosynth Res 123:285–295
Inganas O, Sundstrom (2016) Solar energy for electricity and fuels. Ambio 45 (S1), 15e23 (2016). https://doi.org/10.1007/s13280-015-0729-6
Mueller S, Kwik J (2013) Corn Ethanol: emerging plant energy and environmental technologies, energy resources center. College of Engineering, University of Illinois—Chicago
Mishra A, Kumar A, and Ghosh S(2018) Energy assessment of second generation (2G) ethanol production from wheat straw in Indian scenario, 3 Biotech. Mar; 8(3):142. https://doi.org/10.1007/s13205-018-1135-0
RFA (2016) Re-examining corn ethanol's energy balance ratio ethanolrfa.org › re-examining-corn-ethanols-energy-ba
Scaife MA, Nguyen GT, Rico J, Lambert D, Helliwell KE, Smith AG (2015) Establishing Chlamydomonas reinhardtii as an industrial biotechnology host. Plant J 82:532–546
Shapouri H, Gallagher P, Nefstead W, Schwartz R, Noe S, Conway R (2010) In: 2008 Energy balance for the corn-ethanol industry, agricultural economic Report No 846, Office of Energy Policy and New Uses, US Department of Agriculture
Wang M, Han J, Dunn JB, Cai H, Elgowainy A (2012) Well-to-wheels energy use and greenhouse gas emissions of ethanol from corn, sugarcane and cellulosic biomass for US use. Environ Res Lett 7:1–13. https://doi.org/10.1088/1748-9326/7/4/045905
Yacobucc (2008) CRS report for congress fuel ethanol: background and public policy issues. Updated 24 Apr 2008
Yacobucci B, Womach J (2003) Fuel ethanol: background and public policy issues. Library of Congress. Washington DC. Available at: http://www.ethanol-gec.org/information/briefing/1.pdf

Further Reading

International Energy Agency (IEA) (2020) Bioenergy task 39. Database on facilities for the production of advanced liquid and gaseous biofuels for transport. https://demoplants.bioenergy2020.eu/

RFA—Renewable Fuels Association (2013) Battling for the barrell ethanolrfa.3cdn.net/dc207800043a5aa5aa_y5im6rokb.pdf

Ruan Z, Wang X, Liu Y, Liao W (2019) Corn, integrated processing technologies for food and agricultural by-products. https://doi.org/10.1016/B978-0-12-814138-0.00003-4

Takahashi S, Tanifuji K, Shiell K, Fatehi P, Jahan MS, Ohi H, Ni Y (2013) Removal of acetic acid from spent sulfite liquor using anion exchange resin for effective xylose fermentation with Pichia stipitis. BioResources 8(2):2417–2428

Ugwu CU, Ogbonna J, Tanaka H (2002) Improvement of mass transfer characteristics and productivities of inclined tubular photo bioreactors by installation of internal static mixers. Appl Microbiol Biotechnol 58(5):600–607

von Sivers M, Zacchi G, Olsson L, Hahn-Hägerdal B (1994) Cost analysis of ethanol production from willow using recombinant *Escherichia coli*. Biotechnol Prog 10:555–560

Chapter 6
Pretreatment of Lignocelluloses Biomass for Bioethanol Production

Abstract Different methods—physical, chemical and biological used for pretreatment of lignocelluloses are discussed. These processes are strongly interdependent. Combinatorial pretreatment methods are more effective in improving the biomass digestibility, and are generally used in designing leading pretreatment technologies.

Keywords Biofuels · Bioethanol · Physical pretreatment · Physicochemical pretreatment · Chemical pretreatment · Biological pretreatment

Different methods—physical, chemical and biological methods—are used for pretreatment of lignocelluloses. These processes are strongly interdependent. None of the pretreatment methods are found to be ideal. Obstacles include the following:

Several inhibitory products are produced. These are acids, furans, phenols. Also, high particle load, high energy requirement, and effective separation of soluble sugars from solid residues are required. Different feedstocks require specific pretreatment conditions and mechanistic models could help in the reasonable design of such processes (Zhang 2008; Zhang et al. 2008, 2009a, b; Eggeman and Elander 2005; Cheah et al. 2020). Lignocellulose pretreatment methods need to be optimized as they are one of the most costly steps in the production of bioethanol. Pretreatment cost is about ~30 US cents/gallon of cellulosic ethanol produced (Mosier et al. 2005b). Pretreatment actually changes the biomass macroscopic and microscopic size and structure and also submicroscopic chemical composition and structure so that hydrolysis of carbohydrates to monomeric sugars can be obtained more quickly and with higher yields (Sun and Cheng 2002; Moiser et al. 2005b; Tucker et al. 2003). Pretreatment of cellulosic biomass economically is a major challenge of cellulose to ethanol technology research and development. Native lignocelluloses are very recalcitrant to enzymatic digestion. As a result, several thermochemical pretreatment methods have been developed for improving digestibility (Wyman et al. 2005a, b).

There is a direct correlation between the removal of lignin and hemicellulose on the digestibility of cellulose (Kim and Holtzapple 2006a, b). Thermochemical

Some excerpts taken from Bajpai (2016) Pretreatment of lignocellulosic biomass for biofuel production. Springer Briefs in Molecular Science, Springer Nature

P. Bajpai, *Developments in Bioethanol*, Green Energy and Technology,
https://doi.org/10.1007/978-981-15-8779-5_6

methods are found to be more promising in comparison to biological methods for the conversion of lignin fraction of cellulosic biomass. Actually, lignin has an unfavorable effect on enzyme hydrolysis. Lignin can also serve as a energy source and potential co-products that have important benefits in a life cycle context (Sheehan et al. 2003). Pretreatment is conducted using different methods (Cadoche and López 1989; Gregg and Saddler 1996; Kim et al. 2003; Damaso et al. 2004; Kuhad et al. 1997; Keller et al. 2003; Itoh et al. 2003; Kuo and Lee 2009; Zhao and Liu 2012; Zhang et al. 2009a, b, Zhang et al. 2012a, b, 2013; Li et al. 2012a, b, c; Wang et al., 2011; Zhang et al. 2007a,b; Li et al. 2012; Wang et al. 2011; Zhang et al. 2007) (Table 6.1).

Technoeconomic analysis have been conducted for assessing the cost and performance of pretreatment methods (Eggeman and Elander 2005).

In case of combinatorial pretreatment methods, physical parameters such as pressure or temperature or a biological treatment are combined with chemical treatments and are termed physicochemical or biochemical pretreatment methods. Examples are

- Ammonia fibre/freeze explosion (AFEX)—A good example of a physicochemical method (Sun and Cheng 2002)
- Bioorganosolv—A good example of biochemical method (Itoh et al. 2003).

Table 6.1 Processes for the pretreatment of lignocellulosic feedstocks

Physical Pretreatment *Mechanical comminution* *High energy radiation* *Pyrolysis*
Physicochemical Pretreatment *Steam explosion* *Liquid hot water pretreatment (LHW)* *Ammonia fibre/freeze explosion (AFEX), Ammonia recycle percolation (ARP) and Soaking aqueous ammonia (SAA)* *Carbon dioxide explosion*
Chemical Pretreatment *Oxidative delignification* *Acid Treatment* *Alkali treatment* *Organosolv Process* *Sulphite Pretreatment*
Cellulose solvent-based lignocellulose pretreatments *Ionic liquids* *Aqueous N-Methylmorpholine-N-Oxide* *Urea/Sodium hydroxide* *N, N-Dimethylacetamide (DMAc)/LiCl*
Biological treatment
Other Methods *Glycerol* *Wet Oxidation* *Micorowaves pretreatment* *Ultrasonication*

These combinatorial pretreatment methods are more effective in improving the biomass digestibility, and are generally used in designing leading pretreatment technologies.

6.1 Physical Pretreatment

6.1.1 Mechanical Comminution

This method increases the available specific surface area, and reduces degree of polymerization (DP) and cellulose crystallinity (Sun and Cheng 2002). Different methods have been used for improving the digestibility of lignocellulosic biomass (Palmowski and Muller 1999). These are coarse size reduction, chipping, shredding, grinding and milling (hammer- and ball-milling [wet, dry, vibratory rod/ball milling]), compression milling, ball milling/beating, agitation bead milling, pan milling. Other types of milling methods have been also used (Murnen et al. 2007; Isci et al. 2008, Kim and Lee 2005a, b; Dale and Moreira 1982; Ryu and Lee 1982). These are fluid energy milling, two roll milling, colloid milling. Attrition and disk refining have been also examined for pretreatment (Mes-Hartree et al. 1988; Murnen et al. 2007). Simultaneous ball milling/attrition and enzymatic hydrolysis has been also used for treatment for lignocelluloses (Zheng, et al. 1998; Ben-Ghedalia and Miron 1981). Vibratory ball milling appears to be more effective in comparison to ordinary ball milling for improving the biomass digestibility for pretreating aspen and spruce chips (Neely 1984). Only compression milling process has been studied on a commercial scale (Kilzer and Broido 1965).

Chipping reduces heat and mass transfer limitations. Chipping reduces the biomass size to 10–30 mm. When the particle size of biomass is reduced below 40 mesh (0.400 mm), there is not much effect on the rate and yield of biomass hydrolysis (Chang et al. 1997).

Grinding and milling are more efficient in reducing the particle size and cellulose crystallinity in comparison to chipping. This could be due to the generation of shear forces during milling. Grinding and milling can reduce the particle size to 0.2–2 mm. The type and duration of milling and also the type of biomass determine the increase in specific surface area, final degree of polymerization and the net reduction in cellulose crystallinity. Vibratory ball milling is more effective in comparison to ordinary ball milling in reducing cellulose crystallinity of spruce and aspen chips. Disk milling which produces fibres, is more effective in improving cellulose hydrolysis in comparison to hammer milling which produces fine bundles (Zhua et al. 2009).

The energy requirements of mechanical comminution of lignocelluloses depend on the following factors:

- Characteristics of biomass
- Final particle size required

Table 6.2 Benefits of performing milling after chemical pretreatment

Significantly reduces milling energy consumption
Does not result in the generation of fermentation inhibitors
Reduces cost of solid liquid separation because the pretreated chips can be easily separated
Eliminates energy intensive mixing of pretreatment slurries and liquid to solid ratio

In comparison to agricultural residues, more energy is needed in case of hardwoods (Cadoche and López 1989). Size reduction has been used in most studies of hydrolysis. But not much information is available on the characteristics of the substrate and also the energy consumed during the process (Zhu et al. 2005). Milling process increases bioethanol, biogas and biohydrogen yields (Delgenes et al. 2002). Milling is not economically feasible because it is highly energy intensive (Hendricks and Zeeman 2009). Milling can be performed before or after chemical pretreatment. Several benefits are obtained (Table 6.2) (Zhu et al. 2010a; Zhua et al. 2009):

6.1.2 High Energy Radiation

High energy radiation methods improve the digestibility of cellulosic biomass. These include the following (Yang and Wyman 2008; Youssef and Aziz 1999; Imai et al. 2004; Nitayavardhana et al. 2008; Wojciak and Pekarovicova 2001; Bak et al. 2009; Shin and Sung 2008; Dunlap and Chiang 1980; Kitchaiya et al. 2003; Maa et al. 2009; Saha et al. 2008; Zhu et al. 2005; Yang and Wyman 2008):

- Gamma rays
- Ultrasound
- Electron beam
- Pulsed electrical field
- Ultraviolet rays
- Microwave heating

The action mode behind the high energy radiation could be attributed to

- One or more changes of features of cellulosic biomass including increase of specific surface area
- Reduction of the degrees of polymerization and crystallinity of cellulose
- Hydrolysis of hemicelluloses
- Partial depolymerization of lignin

But, these high energy radiation methods are mostly energy-intensive, slow and expensive (Chang et al. 1981; Lin et al. 1981). They are also strongly substrate-specific (Dunlap and Chiang 1980). So these methods lack commercial appeal based on current estimation of overall cost.

6.1.3 Pyrolysis

Pyrolysis has also been explored for pretreatment of lignocellulosic feedstocks (Kumar et al. 2009). Treatment of materials at temperatures more than 300 °C, decomposes cellulose rapidly and produces gaseous products and residual char (Kilzer and Broido 1965; Shafizadeh and Brad-bury 1979). The decomposition is much slower and less volatile products are generated at reduced temperatures. Mild acid hydrolysis (with 1 N sulfuric acid for 2.5 h at 97 °C) of the residues from pyrolysis pretreatment has shown 80 to 85% conversion of cellulose to reducing sugars with more than 50% glucose (Fan et al. 1987). The process can be improved with oxygen (Shafizadeh and Bradbury 1979). The decomposition of pure cellulose can take place at reduced temperature in the presence of zinc chloride or sodium carbonate catalyst (Shafizadeh and Lai 1975).

Production of transportation fuels from biomass via a so-called biomass-to-liquids route was reported (Zwart et al. 2006). Biomass was converted to syngas from which high quality Fischer-Tropsch fuels are produced. Chipping, pelletization, torrefecation, and pyrolysis have been studied as pretreatment methods for biomass-to-Fischer-Tropsch -fuel conversion. Pretreatment by torrefecation was found to be much better in comparison to pyrolysis

6.2 Physicochemical Pretreatment

6.2.1 Steam Explosion

Steam explosion is mostly used physicochemical method of biomass pretreatment. This process has been researched widely (Agbor et al. 2011). The term "autohydrolysis" has also been used as a synonym for steam explosion. It describes the changes which take place during this process (Chandra et al. 2007; McMillan 1994a, b; Saddler et al. 1993). Biomass is chipped, ground or merely raw preconditioned and then physically pretreated with high pressure saturated steam at pressures between 0.7 and 4.8 MPa and temperatures of about 160–240 °C (Agbor et al. 2011). The pressure is maintained for several seconds to a few minutes for promoting hydrolysis of hemicelluloses and then released. This process causes degradation of hemicelluloses and lignin transformation because of high temperature, thereby increasing the potential of cellulose hydrolysis. Enzymatic hydrolysis was 90% in 24 h for poplar chips pretreated by steam explosion, in comparison to only 15% hydrolysis of untreated chips (Grous et al. 1986). Several factors affect steam explosion pretreatment (Duff and Murray 1996):

- Moisture content
- Temperature
- Residence time
- Size of the chips

Best hemicellulose solubilization and hydrolysis can be achieved by using either high temperature and short residence time (270 °C, 1 min) or lower temperature and longer residence time (190 °C, 10 min) (Duff and Murray 1996). Reduced temperature and longer residence time appear to be promising as the formation of degradation products of sugars which inhibit subsequent fermentation are avoided (Wright 1998).

Hemicellulose is the major fraction of the carbohydrates. These are solubilized in the liquid phase during pretreatment, while lignin gets transformed due to high temperature. The accessibility of cellulose remaining in the solid fraction increases. Because of this, the digestibility of the lignocelluloses increases. Hydrolysis of hemicellulose is thought to be mediated by the acetic acid produced from acetyl groups associated with hemicellulose and other acids produced during pretreatment, which may further catalyse the hydrolysis of hemicellulose forming monomers of glucose and xylose (Mosier et al. 2005b; Weil et al. 1997).

Use of carbon dioxide or sulfur dioxide, sulfuric acid as a catalyst can improve the steam explosion. The use of acid catalyst results in various advantages such as increase in the recovery of hemicellulose sugars, reduction in the production of inhibitory compounds and an increase in the enzymatic hydrolysis on the solid residue (Mosier et al. 2005b; Sun and Cheng 2002).

Steam explosion process is found to be very effective for pretreating agricultural residues and hardwoods, but not found so effective on softwoods (Agbor et al. 2011). The uncatalysed steam pretreatment process has been used in the Masonite process for producing fibre board and other products (Avella and Scoditti 1998; Galbe and Zacchi 2007; Mosier et al. 2005a, b).

6.2.2 Liquid Hot Water Pretreatment (LHW)

This process is similar to steam explosion but in this case water is used at high temperatures instead of steam (Agbor et al. 2011). Following terms have been used to describe LHW:

- Solvolysis
- Hydrothermolysis
- Aqueous fractionation
- Aquasolv

In this process lignin is removed and hemicellulose is hydrolysed and cellulose becomes more accessible (Yang and Wyman 2004). The formation of fermentation inhibitors is avoided which take place at high temperatures. In this process, lower temperatures—optimum between 180 and 190 °C for corn stover and low dry matter-about 1–8% content are used which result in the production of more polysaccharides and oligosaccharides. The temperature of 160–190 °C for pH controlled LHW pretreatment and 170 to 230 °C have been reported depending on the sternness of the pretreatment (Wyman et al. 2005a, b; Bobleter 1994). LHW pretreatment has been performed using co-current, counter current and flow through reactor configuration depending on the direction of the flow of the water and biomass into the reactor. The water and biomass are brought in contact at temperatures of 200–230 °C for about 15 min. The hemiacetal linkages are broken by hot water and acids are released during biomass hydrolysis. This facilitates the breakage of ether linkages in biomass (Antal 1996). The cleavage of O-acetyl groups and uronic acid substitutions on the hemicellulose could help or interfere LHW pretreatment, as the release of these acids helps in catalysing the formation and removal of oligosaccharides, or further hydrolyze hemicellulose to monomeric sugars. These can be later degraded to aldehydes i.e. furfural from pentoses and 5-hydroxymethyl furfural from hexoses which inhibit microbial fermentation (Palmqvist and Hahn-Hägerdal 2000a). The production of monosaccharides and the subsequent degradation products which further catalyze cellulosic hydrolysis during LHW pretreatment can be reduced by maintaining the pH between 4 and 7 (Kohlmann et al. 1995). The flow through reactor configuration in which hot water is passed over a stationary bed of lignocellulose was reported by Yang and Wyman (2004). It was found to be the more effective configuration for removing hemicellulose and lignin at same severity. High lignin solubilization affects recovery of hemicelluloses (Mok and Antal Jr 1992, 1994). Acid catalyst can be used making the process similar to dilute acid pretreatment. But catalytic degradation of sugars results in unwanted side products. During pretreatment, the pKa of water and pH is affected by temperature, therefore potassium hydroxide is used for maintaining the pH above 5 and below 7 for reducing the formation of monosaccharides which are degraded to fermentation inhibitors (Mosier et al. 2005b; Weil et al. 1998).

LHW pretreatment has been studied at laboratory scale. pH controlled LHW pretreatment is considered for pretreatment of corn fibre on a commercial scale. Mosier et al. (2005a) successfully pretreated corn fibre slurry in a 163 L/min reactor with a 20 min residence time. This study showed the possibility of scaling-up LHW pretreatment for pretreatment of large quantities of corn fibre.

6.2.3 Ammonia Fibre/Freeze Explosion (AFEX), Ammonia Recycle Percolation (ARP) and Soaking Aqueous Ammonia (SAA)

These are alkaline pretreatment methods. In these methods liquid ammonia is used for pretreating the biomass (Agbor et al. 2011; Kumar et al. 2009).

In AFEX process, lignocelluloses are treated with liquid ammonia at high temperature and pressure for a certain period of time, and then the pressure is quickly reduced.

In ARP process, aqueous ammonia is passed in the biomass at high temperature, after which ammonia is recovered. AFEX process is similar to steam explosion process operating at high pressure but it is performed at ambient temperatures, whereas ARP is performed at high temperatures.

Modified version of AFEX is SAA process. It uses aqueous ammonia for treating biomass in a batch reactor at moderate temperatures for reducing the liquid throughput during pretreatment (Kim and Lee 2005a, b). At ambient temperatures the duration could be up to 10–60 days whereas at higher temperatures, the effect of ammonia is rapid and the pretreatment time is reduced to minutes (Alizadeh et al. 2005; Kim et al. 2000).

AFEX process is similar to steam explosion. It significantly improves the saccharification rates of several herbaceous crops and grasses. It can be used for the pretreatment of different types of lignocelluloses such as alfalfa, wheat straw, wheat chaff (Mes-Hartree et al. 1988), barley straw, corn stover, rice straw (Vlasenko et al. 1997), bagasse (Holtzapple et al. 1991), coastal Bermuda grass, switchgrass (Reshamwala et al. 1995), municipal solid waste, softwood newspaper, kenaf newspaper (Holtzapple et al. 1992a), aspen chips (Tengerdy and Nagy 1988). Compared to acid pretreatment and acid-catalyzed steam explosion, the AFEX pretreatment does not considerably solubilize hemicellulose (Mes-Hartree et al. 1988; Vlasenko et al. 1997).

Mes-Hartree et al. (1988) made a comparison of steam and ammonia pretreatment for enzymatic hydrolysis of wheat straw, wheat chaff, alfalfa stems and aspen wood. Steam explosion solubilized the hemicellulose, whereas AFEX did not. After AFEX pretreatment, the composition of the materials was found to be the same as the original materials. More than 90% hydrolysis of cellulose and hemicellulose was obtained after AFEX pretreatment of Bermuda grass (approximately 5% lignin) and bagasse (15% lignin) (Holtzapple et al. 1991). But, the AFEX process was not found to be effective for the biomass having high lignin content such as newspaper (18–30% lignin) and aspen chips (25% lignin). Hydrolysis yield of AFEX-pretreated newspaper was 40% and below 50% for aspen chips (McMillan 1994a, b). Ammonia should be recycled after the pretreatment for reducing the cost and protecting the environment. In an ammonia recovery process, a superheated ammonia vapor having a temperature of 200 °C was used for vaporizing and striping the residual ammonia from the pretreated biomass and the evaporated ammonia was removed from the system by a pressure controller for recovery (Holtzapple et al. 1992b). In the ammonia pretreatment, inhibitors are not produced for the downstream biological processes, so water wash is not needed (Dale et al. 1984; Mes-Hartree et al. 1988). Holtzapple et al. (1990) found that AFEX pretreatment does not need small particle size to be effective.

In the AFEX and ARP processes, the lignocelluloses are treated with ammonia at a given temperature and high pressure. This results in swelling and phase change in cellulose crystallinity of biomass and alteration and removal of lignin. The reactivity of the left behind carbohydrates after pretreatment increases. The pretreated biomass can be easily hydrolysed with close to theoretical yields after enzymatic hydrolysis at reduced enzyme dose in comparison to pretreated biomass from other PPMs (Foster et al. 2001; Holtzapple et al. 1991; Kim and Lee 2002; Vlasenko et al. 1997). In AFEX process, biomass is contacted with anhydrous liquid ammonia at a dose of 1–2 kg of ammonia/kg of dry biomass at 60–90 °C for 10–60 min and pressures above 3 MPa. In a closed vessel, the biomass and ammonia mix is heated for about 30 min to the required temperature under pressure. After holding the target temperature for about 5 min, the vent valve is opened quickly to relieve the pressure. The rapid release causes evaporation of the ammonia (volatile at atmospheric pressure) and a concomitant drop in temperature of the system (Alizadeh et al. 2005; Dale and Moreira 1982). The chemical effect of ammonia under pressure causes the biomass to swell, therby increasing the accessible surface area while decrystalizing cellulose. This results in a phase change in the crystal structure of cellulose I to cellulose III (Mosier et al. 2005b; O'Sullivan 1996). A small amount of hemicellulose is solubilized in the oligomeric form during AFEX pretreatment. The lignin distribution in the biomass remains the same after this pretreatment, but the lignin structure is severely changed. This increases the water holding capacity and digestibility. The combined physical and chemical changes substantially increase the susceptibility of the pretreated biomass to subsequent enzymatic hydrolysis (Dale et al. 1984; Dale and Moreira 1982; Galbe and Zacchi 2007; Holtzapple et al. 1991; Kim and Lee 2005a, b). Use of milder process conditions reduces the formation of sugar degradation products and fermentation inhibitors. AFEX process can substantially improve the saccharification rates of herbaceous plants, agricultural residues, and municipal solid waste with about 90% of the theoretical hydrolysis of AFEX pretreated coastal Bermuda grass (Holtzapple et al. 1992a; Mes-hartree et al. 1988).

ARP process uses ammonia percolating in a flow through mode through a packed bed reactor (Agbor et al. 2011). In this process 5–15% aqueous ammonia is passed through biomass in a flow through column reactor at high temperatures (150–180 °C) and a flow rate of 1–5 ml/min with residence time of 10–90 min, after which the ammnonia is recycled or recovered (Kim et al. 2003). Under these conditions, there is simultaneous removal of lignin and hydrolysis of hemicellulose and the reduction in cellulose crystallinity.

Kim et al. (2006), reported a low liquid ARP using one reactor void volume with 15 wt% ammonia in an attempt to optimize the process and establish a continuous process, by dropping the liquid ammonia throughput. Under optimized conditions, the digestibility of the low liquid ARP pretreated biomass was at least 86% of the theoretical maximum even at the lowest enzyme dose of 7.5 FPU/g glucan. Both

AFEX and ARP are effective processes for herbaceous plants, agricultural residues and municipal solid waste. ARP pretreatment is found to be effective on hardwoods also (Iyer et al. 1996; Kim and Lee 2005a, b).

Chan et al. (2014) conducted a carbon dioxide added ammonia explosion pretreatment for ethanol production from rice straw. Using response surface methodology, the pretreatment conditions, such as ammonia concentration, carbon dioxide dose, residence time, and temperature were optimized. The response for optimization was defined as the glucose conversion rate. The optimized conditions resulting in maximal glucose yield (93.6%) were determined as 14.3% of ammonia concentration, 2.2 MPa of carbon dioxide loading level, 165.1 °C of temperature, and 69.8 min of residence time. Scanning electron microscopy analysis showed that pretreatment of rice straw strongly increased the surface area and pore size, which increased enzymatic accessibility for enzymatic saccharification. Finally, an ethanol yield of 97% was obtained via simultaneous saccharification and fermentation. This suggests that carbon dioxide added ammonia pretreatment is a suitable process for bioethanol production from rice straw.

6.2.4 Carbon Dioxide Explosion

This process involves the use of supercritical carbon dioxide under pressure for increasing the digestibility of lignocelluloses (Agbor et al. 2011). Supercritical carbon dioxide explosion, has a lower temperature in cmparison to steam explosion and perhaps has a reduced expenditure as compared to ammonia explosion. Supercritical fluid refers to a fluid which is in a gaseous form but compressed at temperatures above its critical point to a liquid like density. High pressure carbon dioxide, and mainly supercritical carbon dioxide is passed into the reactor and then released by an explosive decompression. As carbon dioxide forms carbonic acid when dissolved in water, the acid increases the hydrolysis rate. Molecules of carbon dioxide are comparable in size to water and ammonia and should be able to penetrate small pores accessible to water and ammonia molecules. Carbon dioxide was found helpful in hydrolyzing hemicellulose and also cellulose. Furthermore, the low temperature prevents any appreciable decomposition of monosaccharides by the acid. Upon a sudden release of the carbon dioxide pressure, the disruption of the cellulosic structure increases the accessible surface area of the substrate to hydrolysis.

This supercritical carbon dioxide pretreatment process does not generate harmful chemicals (Agbor et al. 2011). Therefore it is an environment-friendly method with its own niche in biomass processing. Research has been conducted on few materials such as Avicel cellulose, recycled paper, sugarcane bagasse, aspen, Southern yellow pine, corn stover, switch grass (Zheng et al. 1998; Kim and Hong 2001; Narayanaswamy et al. 2011).

In this process supercritical carbon dioxide is passed to biomass placed in a high pressure vessel (Kim and Hong 2001), or introduced at high pressure (1000–4000 psi) (Zheng et al. 1995). The vessel is heated to the desired temperature and held for a certain period of time, usually several minutes when conducted at high temperatures up to 200 °C (Hendricks and Zeeman 2009). Carbon dioxide enters the biomass at high pressure and it is assumed that once dissolved in water, carbon dioxide will produce carbonic acid which helps in the hydrolysis of hemicellulose. The pressurized gas which is released results in the disruption of the biomass structure. This increases the accessible surface area (Zheng et al. 1995). This process is not effective on biomass with no moisture content but increases the hydrolytic yield of moisture containing biomass with the yield increasing proportionately with the moisture content of the unprocessed feedstock (Kim and Hong 2001). The attractive features of this process are presented below:

- Lower cost
- No generation of toxins
- Use of lower temperatures
- High solids capacity

The high cost of equipment which can tolerate high pressure conditions of carbon dioxide explosion pretreatment is a strong limitation to the application of this process on a commercial scale. Furthermore, the effects on carbohydrate components have yet to be shown.

6.3 Chemical Pretreatment

6.3.1 Oxidative Delignification

Oxidizing agents such as hydrogen peroxide, ozone, oxygen or air can also be used for delignification of lignocelluloses (Bensah and Mensah 2013b; Hammel et al. 2002; Nakamura et al. 2004). In this process, lignin is converted to acids. But, these acids can act as inhibitors during fermentation process. Therefore, these acids have to be removed (Alvira et al. 2010). Also, oxidative treatment damages the hemicelluloses of the lignocellulose complex and a major portion of the hemicellulose gets degraded and becomes unavailable for fermentation (Lucas et al. 2012). The main oxidative delignification process are hydrogen peroxide, ozone, sulphur trioxide.

6.3.1.1 Hydrogen Peroxide

Hydrogen peroxide degrades into hydrogen and oxygen and does not leave residues in the biomass and is the most commonly used oxidizing agentt (Uppal et al. 2011). Process variations that have produced high sugar yields include the addition of catalysts such as manganese acetate, post-pretreatment acid saccharification, and alkaline-peroxide application without post-pretreatment washing (Lucas et al. 2012; Uppal et al. 2011; Banerjee et al. 2012). Removal of lignin from lignocelluloses leads to the exposure of cellulose and hemicellulose causing enhanced enzymatic hydrolysis (Hammel et al. 2002). The yield of enzymatic hydrolysis followed can be as high as 95%. Combined hydrogen peroxide and biological treatment of rice hull was performed by Yu et al. (2009). The combined pretreatments led to significant increases of the lignin degradation. Under optimized conditions, hydrogen peroxide (2%, 48 h) and P. ostreatus (18 days) led to 39.8–9.6% of net yields of total sugar and glucose, respectively. It was about 5.8 times and 6.5 times more in comparison to the fungal pretreatment (18 days) alone, and was also slightly higher as compared to fungal pretreatment alone for 60 days.

6.3.1.2 Ozonolysis

Ozonolysis can be used for reducing the lignin content of lignocelluloses. This increases the in vitro digestibility of the treated material, and does not generate toxic residues unlike other chemical treatments. Ozone can be used for degrading lignin and hemicelluloses in several materials such as poplar saw dust, wheat straw, bagasse, green hay, peanut, pine and cotton straw (Ben-Ghedalia and Miron 1981; Neely 1984; Vidal and Molinier 1988). The degradation was mostly limited to lignin; hemicellulose was slightly affected, whereas cellulose was hardly affected. The rate of enzymatic hydrolysis increased by 5 times following 60% removal of the lignin from wheat straw in ozone pretreatment (Vidal and Molinier 1988). Enzymatic hydrolysis yield increased from 0 to 57% as the percentage of lignin reduced from 29 to 8% after ozonolysis of poplar sawdust (Vidal and Molinier 1988).

6.3.1.3 Sulfur Trioxide

Yao et al. (2011) examined the use of sulfur trioxide in a process called sulfur trioxide microthermal explosion (STEX) to pretreat lignocelluloses such as rice straw. Biomass is hanged above a solution of oleum and sodium hydroxide (1% w/v) and swirled in a test tube at 50 °C/1 atm for 7 h, followed by washing to obtain the solids. The internal explosion is believed to take place because of heat released from sulfur trioxide–straw reaction which causes rapid expansion of air and water from the interior of the biomass which results in increased structural changes and pore volume and partial removal of hemicellulose and lignin (Yao et al. 2011; Li et al. 2012a). Pretreatment of rice straw at the above-mentioned conditions resulted in the saccharification yield of 91% (Yao et al. 2011). The effective handling of sulfur trioxide will be a major challenge because of its corrosiveness regarding this emerging pretreatment.

6.3.2 Acid Treatment

Concentrated acids (sulfuric acid and hydrochloric acid) have been used to treat lignocellulosic materials (Agbor et al. 2011; Bensah and Mensah 2013a; Chaturvedi and Verma 2013; Kumar et al. 2009). Although they are powerful agents for cellulose hydrolysis, concentrated acids are toxic, corrosive, and hazardous and require reactors that are resistant to corrosion. To make the process economically feasible, the concentrated acid must be recovered after hydrolysis (Sivers and Zacchi 1995). Dilute acid hydrolysis has been successfully used for pretreatment of lignocellulosic materials. It has been used in case of wide range of feedstocks, including softwood, hardwood, herbaceous crops, agricultural residues, wastepaper, and municipal solid waste. It performed well on most biomass materials. Hydrochloric acid, nitric acid, phosphoric acid, and sulfuric acid have been examined for use in biomass pretreatment. Dilute sulfuric acid is commonly used as the acid of choice (Kim et al. 2000; Mosier et al. 2005b; Nguyen et al. 2000; Torget et al. 1992). It is mixed with biomass to solubilize hemicellulose thereby increasing the accessibility of the cellulose in the biomass. This mixture can be heated directly with the use of steam as in steam pretreatment, or indirectly via the vessel walls of the reactor. The dilute sulfuric acid pretreatment can significantly improve cellulose hydrolysis and can obtain high reaction rates (Esteghalian et al. 1992). At moderate temperature, direct saccharification suffered from low yields because of sugar decomposition. High temperature in dilute acid treatment is favorable for cellulose hydrolysis (McMillan 1994a, b). Recently developed dilute acid hydrolysis processes use less-severe conditions and achieve high xylan to xylose conversion yields. Achieving high xylan to xylose conversion yields is necessary to achieve favorable overall process economics because xylan accounts for up to a third of the total carbohydrate in many lignocellulosic materials (Hinman et al. 1992).

6.3.3 Alkali Treatment

Alkaline pretreatment is one of major chemical pretreatment technologies. It uses various bases, including sodium hydroxide, potassium hydroxide, calcium hydroxide (lime), aqueous ammonia, ammonia hydroxide and sodium hydroxide in combination with hydrogen peroxide or others. Pretreatment with alkali cause swelling of biomass, which increases the internal surface area of the biomass, and decreases both the degree of polymerization, and cellulose crystallinity. Alkaline pretreatment disrupts the lignin structure and breaks the linkage between lignin and the other carbohydrate fractions in lignocellulosic biomass, thus making the carbohydrates in the heteromatrix more accessible. The reactivity of remaining polysaccharides increases as the lignin is removed. Acetyl and other uronic acid substitutions on hemicellulose that reduce the accessibility of enzymes to cellulose surface are also removed by alkali pretreatments (Chandra et al. 2007; Chang and Holtzapple 2000; Galbe and Zacchi 2007; Mosier et al. 2005b). However, most of the alkali is utilized. The effectiveness of alkaline pretreatment varies, depending on the substrate and treatment conditions. In general, alkaline pretreatment is more effective on hardwood, herbaceous crops, and agricultural residues with low lignin content than on softwood with high lignin content (Bjerre 1996). The digestibility of sodium hydroxide-treated hardwood increased from 14 to 55% with the decrease of lignin content from 24–55 to 20%. However, no effect of dilute sodium hydroxide pretreatment was observed for softwoods with lignin content higher than 26% (Millet et al. 1976). Dilute sodium hydroxide pretreatment was also effective for the hydrolysis of straws with a relatively low lignin content of 10–18% (Bjerre 1996).

6.3.4 Organosolv Process

Organosolv process has been extensively utilized for extraction of high-quality lignin, which is a value added product, and once the lignin is removed from the biomass, the cellulose fibers become accessible to enzymes for hydrolysis and absorption of cellulolytic enzymes to lignin is minimized, which leads to higher conversion of biomass (Agbor et al. 2011).

The organosolv pretreatment process involves the addition of an aqueous organic solvent mixture to the biomass under specific temperatures and pressures with/without a catalyst (Sun and Cheng 2002; Alriols et al. 2009; Chum et al. 1985). The catalyst used are acid, a base, or a salt and the solvents which are most commonly used are ethanol, methanol, ethylene glycol and acetone (Ichwan and Son 2011). Temperatures used in the process can be as high as 200 °C, but lower temperatures can also be used depending on the type of biomass and the use of a catalyst. The process produces three main fractions—a high purity lignin, a relatively pure cellulose fraction and a hemicellulosic syrup containing C5 and C6 sugars. The pretreated solid residues are separated by filtration and washed with distilled water in order to remove solvents and degradation products which may inhibit downstream process such as enzymatic hydrolysis and fermentation. Pretreatment conditions lead to simultaneous hydrolysis and lignin removal (Conde-Mej´ia et al. 2012; Chum et al. 1985). The presence of acids as a catalyst causes acid-catalyzed degradation of the monosaccharides into furfural and 5-hydroxymethyl furfural followed by condensation reactions between lignin and these reactive aldehydes (Chum et al. 1985). After removal of lignin, the biomass is used for enzymatic hydrolysis (Zhao et al. 2009).

Process variables affect the physical characteristics of the pretreated substrates. In most cases, high temperatures and acid concentrations and also long reaction times cause considerable degradation of sugar into fermentation inhibitors. Sulphuric acid has been used extensively as a catalyst (Park et al. 2010).

Organic solvents/acids that have been used as catalysts include oxalic, formic, acetylsalicylic, salicylic acid, ethanol, methanol, acetone, ethylene glycol, triethylene glycol, and tetrahydrofurfuryl alcohol (Sun and Cheng 2002; Haverty et al. 2012; Mesa et al. 2011; Zhao and Liu 2012).

The Battelle organosolv method involves the use of a ternary mixture of phenol, water, and hydrochloric acid to fractionate the biomass at about 100 °C and at 1 atm (Villaverde et al. 2010). The acid depolymerizes lignin and hydrolyses the hemicellulose fraction and the lignin dissolves in the organic phase (phenol), while the monosaccharides are accumulated in the aqueous phase upon cooling of the fractionated biomass. Similarly, the formic acid organosolv process (formasolv) involves the application of formic acid, water, and hydrochloric acid to depolymerize, oxidize, and dissolve lignin, hemicellulose, and extractives in the biomass, and the precipitation of the lignin is obtained by the addition of water (Villaverde et al. 2010). Formic acid has a good lignin solvency and the process can be conducted under low temperatures and at atmospheric pressure (Zhao and Liu 2012).

Kupiainen et al. (2012) used formic acid on delignified wheat straw pulp at 180–220 °C, yielding a maximum glucose yield of 40%. Organosolv pretreatment with acetic acid produces higher yields than formic since less material is dissolved for a given time. Higher cellulose viscosity in smaller time periods is obtained in acetosolv process (Villaverde et al. 2010).

The use of ethanol in organosolv pretreatment enables the recovery of high value products which include cellulose, sulphur- and chlorine-free lignin, enriched hemicelluloses, and extractives. Via the use of solvents such as ionic liquids, further purification may be obtained (Prado et al. 2012). The ethanosolv process is usually operated under higher pressures and temperatures unlike formasolv process. In addition, reprecipitation of lignin takes place due to lower lignin solubility (Zhao and Liu 2012).The reduced toxicity of ethanol compared to solvents such as methanol to the downstream fermentation process and the fact that ethanol is the final product are additional benefits (Chum et al. 1985; Kim et al. 2011). Generally, lower ethanol/water ratios favour hemicellulose hydrolysis and enzymatic degradability of pretreated biomass since ethanol inhibits the performance of hydrolytic enzymes (Huijgen et al. 2008). Ethanosolv process has been explored for the development of proprietary technologies such as the Alcell process and the Lignol process. Alcell process is a sustainable alternative to kraft pulping (Pye and Lora 1991) and the Lignol process—a biorefinery platform that uses aqueous ethanol (50%w/w) for pretreating lignocellulosic biomass at 200 °C and 400 psi to separate the different components in biomass (Arato et al. 2005; Pan et al. 2005). High sugar yields and product recovery have been observed in case of ethanosolv pretreatment of various materials including hybrid poplar (Pan et al. 2006a) and Japanese cypress (Hideno et al. 2013). A main advantage is the potential to recover much of the ethanol (Koo et al. 2012; Alriols et al. 2010) and water (Alriols et al. 2009). This reduces the operating cost. Ethanosolv coproducts such as hemicellulose syrup and lignin can serve as feedstocks for the production of high value products. Furthermore, ethanosolv lignins whose molecular weight and functional groups depend on process conditions are known to possess antioxidant properties (Pan et al. 2006b). In several situations, presoaking materials, for example, in acidic medium (Kim et al. 2012) or in bioslurry (Brosse et al. 2010), positively affect the process in terms of sugar yields and lignin removal, among 7 others. Other variations involve the combined use of acid and basic catalysts, microwave-assisted organosolv, and the avoidance of catalysts (Mesa et al. 2011; Li et al. 2012b; Wang et al. 2012). In another variation, the biomass is treated with the inclusion of ferric sulphate and sodium hydroxide to the biomass/liquor (formic acid and hydrogen peroxide) mixture. Formic acid reacts with hydrogen peroxide to form peroxyformic acid and its application in organosolv pretreatment (Milox) of biomass produced good results on hardwoods and also softwoods (Villaverde et al. 2010).

6.3.5 Sulphite Pretreatment

Sulphite application is emerging as a promising pretreatment method due to positive results obtained from several materials (Bensah and Mensah 2013b). University of Wisconsin, Madison researchers have developed an improved pretreatment process for conversion of biomass. This process, known as Sulfite Pretreatment to Overcome Recalcitrance of Lignocellulose (SPORL), reduces the energy consumption needed for size-reduction processes, required before enzymatic hydrolysis, by more than tenfold. The new method can use several aqueous sulfite or bisulfite solutions over a wide range of pH values and temperatures to weaken the chemical structure of the plant material. The improved SPORL method is flexible and can be integrated easily into current pretreatment systems. It is particularly suitable for woody biomass, softwoods such as pines and other conifers and hardwoods such as poplar, willow and eucalyptus. The pH of the pretreatment liquor can be adjusted by reagent, making SPORL easily incorporated into current dilute acid approaches to improve the efficiency of the pretreatment. Depending on the stock material, mechanical size reduction steps such as disk or hammer milling can be implemented directly before or after SPORL. In addition, the final enzymatic hydrolysis can be coupled directly after the pretreatment with or without washing the material or adding a surfactant to aid in the process. The pretreatment also can be used with steam explosion, using bisulfite as a catalyst. The hydrolyzed biomass can be separated and the sugars fermented or catalytically converted into fuels after pretreatment and the sulfonated lignin byproducts can be sold and other wastes burned to produce energy for the process.

The SPORL method is a superior method of biomass pretreatment due to its versatility, efficiency and simplicity. It shows excellent scalability to commercial production. Through decreased size-reduction energy requirements and maximized enzymatic cellulose conversion in a short period of time, this method will increase the energy efficiency of ethanol fermentation and catalytic fuel production processes. This increase in efficiency will allow biofuels and other bioproducts to become economically competitive with petroleum derived fuels and products.

6.4 Cellulose Solvent-Based Lignocellulose Pretreatments

Cellulose solvent-based lignocellulose pretreatments are gaining attention because they can break recalcitrant structure of biomass by increasing cellulose accessibility in a more effective manner than traditional biomass pretreatments (Sathitsuksanoh et al. 2012a). The hydrolysis rate and digestibility of pretreated biomass are increased as a result of this and enzyme use is decreased (Sathitsuksanoh et al. 2009, 2010, 2012a, b). Also, cellulose solvent-based pretreatments may be regarded as a biomass-independent pretreatment. A few cellulose solvent-based strategies are being developed, such as concentrated phosphoric acid (85% (w/w)), ionic liquids, NMMO, NaOH/urea, and DMAc/LiCl.

Cellulose solvent- and organic solvent-based lignocellulose fractionation (COSLIF) was developed to fractionate lignocellulose using a combination of concentrated phosphoric acid as a cellulose solvent and an organic solvent such as acetone or ethanol under modest reaction conditions (Ladisch et al. 1978).The ideas of COSLIF are following (Zhang et al. 2006, Zhang et al. 2007a, b; Zhang and Lynd 2006a, b):

- Removal of partial lignin and hemicellulose—eliminating the major obstacles to hydrolysis and allowing cellulase to access the substrate more efficiently
- De-crystallization of cellulose fibers. providing better cellulose accessibility to cellulase
- Modest reaction conditions—a reduction in sugar degradation, reduced formation of inhibitors, reduced utility consumption, and capital investment

Concentrated phosphoric acid can completely dissolve cellulose fibers. This results in effective disruption of highly ordered hydrogen bonding network of crystalline cellulose (Sathitsuksanoh et al. 2011; Moxley et al. 2008) and drastic increases in CAC (Rollin et al. 2011; Zhu et al. 2009a, b). COSLIF has been demonstrated to efficiently pretreat a variety of feedstocks (Sathitsuksanoh et al. 2009, 2010, 2012a, b; Zhu et al. 2009a, b; Zhang et al. 2007c; Moxley et al. 2008; Conte et al. 2009; Ge et al. 2012).

6.4.1 Ionic Liquids

Novel pretreatment methods using ionic liquids (ILs) are creating new opportunities in the area of biomass conversion (Socha et al. 2014; Sathitsuksanoh et al. 2012a; Wu et al. 2004, 2011). ILs are salts that exist as liquids and are composed entirely of paired ions and have tuneable properties. These are being extensively studied as "green solvents" for several industrial and research applications, particularly in the area of catalysis, chemical synthesis and separation of cellulose-based biomass (Pu et al. 2007). They are able to dissolve large amounts of cellulose at extremely mild conditions. The possibility of recovering almost 100% of the used ILs to their initial purity makes them attractive (Heinze et al. 2005). This technology was used for direct dissolution of cellulose in the commercial Lyocell process as a modern industrial fiber-making (Fink et al. 2001).

ILs convert carbohydrates in lignocellulosic materials into fermentable sugars via two main pathways (Wang et al. 2011b). One is the pretreatment of the biomass to improve its efficiency of enzymatic hydrolysis, and the other deals with the transformation of the hydrolysis process from a heterogeneous to a homogeneous reaction system by dissolution in the solvent.

ILs are recognised to facilitate more green applications in reactions and separations due to their special beneficial properties, such as high thermal stability and negligible vapour pressure. Their very low vapour pressure reduces the risk of exposure which is a clear advantage over the use of the traditional volatile solvents. ILs are increasingly being used to dissolve various lignocellulosic biomass.

Most ILs used in biomass fractionation are imidazonium salts. Studies have shown that 1-allyl-3-methylimidazonium chloride (AMIMCl) and 1-butyl-3-methylimidazonium chloride (BMIMCl) can be used effectively as a non-derivatising solvent for the dissolution of cellulose at temperatures below 100 °C (Zhang and Lynd 2006; Zhu et al. 2006). The cellulose fraction can be recovered by the addition of water, ethanol or acetone. The solvent can be recovered and reused by using various techniques such as reverse osmosis, pervaporation, salting out and ionic exchange.

Ionic liquids are able to dissolve a variety of biomass with different hardness and are introduced as selective solvents of lignin and cellulose (Cao et al. 2010). In this process, biomass is solubilized in the solvent at 90–130 °C and ambient pressure, followed by the addition of water to precipitate the biomass. The process is completed by washing the precipitate. The structure of lignin and hemicellulose remain unchanged after treatment with ionic liquids. This allows the selective extraction of unaltered lignin because lignin is highly soluble in solvents whereas the solubility of cellulose is low (Zhang et al. 2012a, b). This can result in the separation of lignin and the enhancement in cellulose accessibility under ambient temperature and pressure without alkaline or acidic reagents and the formation of inhibitors. These solvents are expensive. But, the cost of their recovery is not high because of the low vapor pressure of the solvents. Nevertheless, cellulase enzyme is irreversibly inactivated in ionic liquid solvents (Zhi et al. 2012), which reduces the biomass conversion efficiency and increases the overall cost. This result shows the need to develop solvents in which cellulase and microorganisms are active.

Application of ionic liquids has opened new ways for the effective utilization of lignocellulosic materials in areas such as biomass pretreatment and fractionation.

6.4.2 Aqueous N-Methylmorpholine-N-Oxide

Aqueous N-Methylmorpholine-N-Oxide (NMMO) has attracted interest for use as a pretreatment solvent. It is a well-known industrial solvent used in the Lyocell process for the production of fibres. Cellulose dissolves without derivatization in

NMMO/H_2O system and the hydrogen bonds in cellulose are broken down in favour of new bonds between cellulose and solvent molecules (Zhao et al. 2007a, b). This leads to swelling and increased porosity, and also reduced degree of polymerization and crystallinity which improves enzymatic saccharification. Addition of boiled distilled water to pretreatment slurry containing dissolved biomass causes cellulose I to precipitate into cellulose II which is more reactive. Regenerated solids are filtered and washed with warm/boiling water until the filtrate is clear. Like ionic liquids, NMMO dissolves biomass with no/less chemical modification at low/moderate temperatures (80–130 °C). Additional favourable characteristics of NMMO pretreatment include high sugar yields, high solvent recovery, formation of low degradation products and no harmful effect on the environment. Further, cellulase activity is not negatively affected by low concentrations (15–20% w/w) of NMMO, indicating the potential of application in continuous processes (Li et al. 2012b).

Aqueous NMMO has already being used on several types of biomass as a sole pretreatment method or in combination with others. A technoeconomic analysis of NMMO pretreatment of spruce for ethanol and biogas production was conducted by Shafiei et al. (2011). They observed relatively high process energy efficiency of 79%. They observed relatively high process energy efficiency of 79%.

6.4.3 Urea/Sodium Hydroxide

The Urea/Sodium hydroxide solutions have been found to dissolve cellulose at a subzero temperature for the homogeneous synthesis of cellulose derivatives (Khodaverdi et al. 2012; Wang et al. 2008, 2011a; Ruan et al. 2004). Khodaverdi et al. (2012) used Urea/Sodium hydroxide solution to pretreat spruce. Spruce showed slight removal of cellulose, hemicellulose, and lignin. However, a significant increase in enzymatic glucan digestibility was observed (Khodaverdi et al. 2012). But, it may be too expensive to prepare prechilled Urea/Sodium hydroxide and recycle this solution, particularly in the case of biomass pretreatment that is used to produce low-value biocommodities. For example, sodium hydroxide-based pulping used to cause serious water pollution in China. So it has been abandoned (Sathitsuksanoh et al. 2012a).

6.4.4 N,N-Dimethylacetamide (DMAc)/LiCl

N,N-Dimethylacetamide (DMAc)/LiCl solution can dissolve cellulose (Cai et al. 2004) because hydrogen bonding of the hydroxyl protons of cellulose with the chloride ions allows the solvent to penetrate into cellulose fibers. DMAc/LiCl is suitable for processing and derivatizing pure cellulose (Striegel 1997). Wang et al. (2011b) conducted a comparative study using different cellulose solvents – LiOH/urea, LiCl/DMAc, concentrated phosphoric acid, NMMO and 1-butyl-3-methylimidazolium chloride. Except for the cellulosic sample regenerated from

LiCl/DMAc system, all the other treated samples showed reduced cellulose crystallinity and degree of polymerization and as a consequence, showed a significant improvement of enzymatic hydrolysis kinetics. The regenerated cellulose from concentrated phosphoric acid almost completely consisted of cellulose II, and obtained the highest saccharification yield (Sathitsuksanoh et al. 2012a).

6.5 Biological Treatment

Biological pretreatment uses wood degrading microorganisms which include white-rot fungi, brown-rot fungi, soft-rot fungi, and bacteria to modify the structure/chemical composition of the lignocellulosic biomass so that the modified biomass is more responsive to enzymatic digestion. Fungi have distinct degradation characteristics on lignocellulosic biomass. Generally, brown and soft rots mainly attack cellulose and impart minor modifications to lignin, whereas white-rot fungi more actively degrade the lignin component. Currently, research is aimed towards finding those organisms which can degrade lignin more effectively and more specifically. White-rot fungi are considered the most promising basidiomycetes for bio-pretreatment of biomass and are the most studied biomass degrading microorganisms (Lee 1997; Sun and Cheng 2002). Brown-rot, white and soft-rot fungi attack wood via the production of enzymes such as lignin peroxidases, polyphenol oxidases, maganesse-dependent peroxidases, and laccases that degrade the lignin. Hatakka et al. (1993) reported the selective delignification of wood and wheat straw by selected white-rot fungi such as; Phanerochaete chrysosporium, Phlebia radiata, Dichmitus squalens, Rigidosporus lignosus, and Jungua separabilima. Lignin depolymerization by these fungi takes several weeks to achieve significant result but can be very selective and efficient (Hatakka 1994; Hatakka et al. 1993).

Hwang et al. (2008) investigated biological pretreatment of wood chips using four different white-rot fungi for 30 days. They observed that the glucose yield of pretreated wood by Trametes versicolor MrP 1 reached 45% by enzymatic hydrolysis whereas 35% solid was converted to glucose during fungal incubation. A Japanese red pine Pinus densiflora (softwood) was pretreated biologically by several white-rot fungi—Ceriporia lacerata, Stereum hirsutum, and Polyporus brumalis and it was found that S. hirsutum is the most effective to degrade lignin and improve the enzymatic digestibility of wood (Lee et al. 2007).

Preliminary studies conducted by Keller et al. (2003) showed a 3- to 5-fold improvement in enzymatic digestibility of corn stover after pretreatment with Cyathus stercoreus; and a 10- to 100-fold reduction in shear force required to obtain the same shear rate of 3.2 to 7.0 rev/s, respectively, after pretreatment with Phanerochaete chrysosporium.

Zhang et al. (2007a, b) screened 35 isolates of white-rot fungi for the biological pretreatment of bamboo for enzymatic saccharification. They observed that Echinodontium taxodii 2538, Trametes versicolor G20 and Coriolus versicolor B1 were the most promising white-rot fungi for significant improvement of enzymatic saccharification and highly selective lignin degradation.

The biological pretreatment appears to be a promising technique and has very evident advantages, including no chemical requirement, low energy input, mild environmental conditions, low energy input and environmentally friendly working manner (Sun and Cheng 2002). However, biological pretreatment is very slow and large amount of space is required to perform this treatment. It also requires careful control of growth conditions (Chandra et al. 2007). In addition, most lignolytic fungi solubilize/consume not only lignin but also hemicellulose and cellulose (Lee et al. 2007; Singh et al. 2008; Kuhar et al. 2008; Shi et al. 2008). Therefore, the biological pretreatment faces technoeconomic challenges and is less attractive commercially.

6.6 Other Methods

6.6.1 Glycerol

Crude glycerol has been also studied as a solvent for fractionating biomass in order to improve the economics of cellulosic ethanol and also upstream biodiesel production. Glycerol pretreatment causes very effective delignification of biomass. Guragain et al. (2011) found the optimum conditions for the use of crude glycerol (water:glycerol = 1:1) in pretreating wheat straw and water hyacinth. Best results were obtained at a temperature of 230 °C for 4 h for wheat straw and 230 °C for 1 h for water hyacinth. Enzymatic hydrolysis of pretreated wheat straw produced reducing sugar yields (mg/g of sample) of 423 and 487 for crude and pure glycerol, respectively, compared to 223 for dilute acid. In addition, hydrolysis tests on water hyacinth gave yields of 705, 719, and 714 for crude glycerol, pure glycerol, and dilute acid, respectively. Ungurean et al. (2011) conducted glycerol pretreatment of different types of wood—poplar, acacia, oak, and fir—and obtained higher cellulose conversion rates compared to dilute acid application. However, combinations of glycerol and acid/IL pretreatment produced higher sugar levels compared to glycerol pretreatment alone. There are wide variations in the composition of crude glycerol which usually contains methanol, ash, soap, catalysts, salts, and nonglycerol organic matter, among others, in different proportions (Yang et al. 2012). While the potential for exploring crude glycerol application together with other methods is high, there is a need to assess the quality of crude glycerol and its effects on sugar and ethanol yields of promising feedstocks.

6.6.2 Wet Oxidation

Wet oxidation process was originally used as a means of wastewater treatment and soil remediation, and later on it was also used in the pretreatment of lignocellulosic feedstocks (Chaturvedi and Verma 2013). Wet oxidation involves water and air, oxygen, or hydrogen peroxide at high pressure and temperature (Varga et al. 2003). Most of the hemicellulose is dissolved in the pretreatment, and it is possible to obtain a moderate to high degree of delignification through oxidation (controlled combustion). In this process, large number of organic polymers mostly hemicellulose and lignin are converted to oxidized compounds, such as low-molecular-weight carboxylic acids, alcohol, or even carbon dioxide and water. Wet oxidization is mostly carried out with the addition of an alkali such as sodium carbonate to reduce the reaction temperature and the amount of hemicellulose being oxidized. The use of hydrogen peroxide has attracted much interest in the recent years, but the high cost of this chemical could make this technology economically prohibitive. The introduction of pure oxygen to the reaction has been challenged, since uncontrolled combustion can occur at the oxygen injection points. Therefore, it is very much unlikely that this pretreatment method will find practical applications in biomass processing. In the wet oxidation process, oxidation of organic matter in the presence of oxygen takes place. When the process takes place at low temperatures, hydrolysis of lignocellulose occurs. At high temperatures, oxidation of lignocellulose occurs with liberation of carbon dioxide and water. Wet oxidation process solubilizes hemicellulose and lignin is degraded into carbon dioxide, water, and carboxylic acids (succinic acid, formic acid, acetic acid, glycolic acid) and phenolic compounds (Bjerre et al. 1996). It was presented as an alternative to steam explosion process which had become the most widely used pretreatment method (Palonen et al. 2004). Industrially, wet air oxidation processes have been used for the treatment of wastes with a high organic matter by oxidation of soluble or suspended materials using either an oxidizing agent such as hydrogen peroxide or oxygen in aqueous phase at high temperatures (180–200 °C) (Jorgensen et al. 2007). The effect of wet oxidation to improve anaerobic biodegradability and methane yields from different biowastes such as food waste, yard waste, and digested biowaste by using thermal wet oxidation was studied by Lissens et al. (2004). Methane yields for wet oxidized yard waste, raw food waste, wet oxidized food, and raw yard waste were 345, 685, 536, 571, and 345 mL of methane/g of volatile suspended solids, respectively. Higher oxygen pressure during wet oxidation of digested biowaste substantially increased the total methane yield and digestion kinetics and also allowed lignin utilization during a subsequent second digestion. The increase of the specific methane yield for the full-scale biogas plant by using thermal wet oxidation was 35–40%, showing that there is still a significant amount of methane that can be harvested from anaerobic digested biowaste. Szijártó et al. (2009) treated common reed (Phragmites australis) by using the wet oxidation process to improve the enzymatic digestibility of reed cellulose to soluble sugars, thus improving the convertibility of reed to ethanol. The most effective treatment increased the digestibility of reed cellulose by cellulase enzymes more than three times compared to the untreated control. About 51.7% of the hemicellulose and 58.3% of the lignin was solubilized and about 87.1% of the cellulose remained in the solids during this wet oxidation process. The conversion of cellulose to glucose was 82.4% after enzymatic hydrolysis of pretreated fibers from the same treatment. Simultaneous saccharification and fermentation of pretreated solids resulted in a final ethanol concentration as high as 8.7 g/L which was 73% of the theoretical yield. In the wet oxidation process, lower amounts of furfural and 5-hydroxymethylfurfural are produced. These are strong inhibitors in the fermentation step, as a result of oxygen degradation of these components. However, for the same reason, the large amount of hemicellulose sugars are lost, and therefore the overall process yield would be reduced and the economy is not attractive. Another concern associated with the oxidation process is its exothermal nature, which requires careful control of the process parameters.

6.6.3 Micorowaves Pretreatment

Microwaves have been used in the treatment of lignocellulosic biomass (Chaturvedi and Verma 2013). Microwaves cause localized heating of biomass which leads to disruption of lignocellulose architecture. Thus, cellulose and hemicellulose become accessible to enzymatic hydrolysis (Sarkar et al. 2012). Microwave chemistry offers many advantages over conventional methods of heating. The process is more energy efficient because microwave irradiation heats the whole volume of a sample whereas conventional heating heats the sample in contact with the reaction vessel before the bulk. The heating effect is almost rapid, unlike conventional heating methods. There is no time spent waiting for the source to heat up or cool down. Microwave reactor operating systems allow easy control of temperature and pressure to exact and steady values. The speed of microwave systems also allows for quick screening of parameters and evaluation of new methods (Kappe 2004). The application of microwave technology in high throughput reactions such as flow reactors adds further to the advantages of microwave irradiation (Roberts and Strauss 2005; Wilson et al. 2004). Su et al. (2010) studied the effects of microwave treatment on sorghum liquor waste for bioethanol production. Their results showed that reducing sugar yield following microwave treatment was very high as compared to untreated waste. Liu et al. (2010) studied microwave pretreatment of recalcitrant softwood in the presence of aqueous glycerol and different organic and inorganic acids. They found that the pulp obtained by organosolvlysis with 0.1% hydrochloric acid (pKa-6) at 180 °C for 6 min gave the highest sugar yield, 53.1%. With other acids such as malonic acid, and phosphoric acid having lower pKa value, a lower sugar yield was obtained. Microwave-assisted glycerolysis is a suitable process for treatment of different types of soft woods.

6.6.4 Ultrasonication

This pretreatment process is based on the principle of cavitation through the use of ultrasonic radiation (Baruah et al. 2018). The shear forces are generated which cleaves the complex network structure of LCB and helps the extraction of cellulose, hemicellulose, and or lignin (Ravindran and Jaiswal 2016). He et al. (2017) examined ultrasound pretreatment on the structural changes of eucalyptus. They observed that the crystallinity of the pretreated wood increased from 34.7 to 35.3% in aqueous soda solution, 32.6–35.5% in distilled water and 33.4–35.5% in aqueous acetic acid. FTIR analysis showed that the increased crystallinity could be due to the effective removal of the amorphous hemicellulose and lignin fractions. Selection of solvents (dilute aqueous solutions of inorganic acids or alkalis, organic solvents or ionic liquids) is crucial in determining the optimum conditions for ultrasonication (Koutsianitis et al. 2015). Several factors affect the sonication treatment. These include the following:

- Ultrasound frequency
- Sonication duration
- Sonication power
- Temperature

Liyakathali et al. (2016) reported that the enzymatic digestibility of bagasse increased with increase in the sonication time and temperature whereas ultrasonic frequencies did not effect enzymatic digestibility.

The use of ultrasonic pretreatment on wood chips for bio-oil production was examined by Cherpozat et al. (2017). The experiments were performed under different conditions of frequency (40, 68, and 170 kHz), treatment time (0.5, 1 and 1.5 h) and power (125, 250, 500, and 1,000 W). Combination of 170 kHz for 0.5 h and 40 kHz for 1.5 h and a power of 1,000 was found to be sufficient. This resulted in 12% increased yield of bio-oil in comparison to untreated wood. Ultrasonication for a prolonged period causes harmful effect because of collision and aggregation between the particles (Ivetić et al. 2017). Similarly, use of high sonication power leads to cavitation near the ultrasound transducer tip. This prohibits the energy transfer to the liquid medium (Subhedar and Gogate 2014).

Another study on the ultrasound pretreatment on sugar beet shreds followed by enzymatic hydrolysis showed a yield of 780 mg/g cellulose, which was 3.7 times higher in comparison to that obtained with untreated samples (Ivetić et al. 2017).

Luzzi et al. (2017) studied the ultrasonication treatment and enzymatic hydrolysis in a single step. The cellulase activity was monitored in terms of temperature, pH, and ultrasound power. Maximum yield of 15.5 UPFml − 1 was obtained at 40 °C, pH 4.6 and power of 44 W. The optimum conditions showed effective sugar production on enzymatic hydrolysis of filter paper and bagasse malt. Nakashima et al. (2016) studied the impact of LCB characteristics, reactor geometry and kinetics during ultrasonic pretreatment (Nakashima et al. 2016).

Ultrasonication appears to be a viable pretreatment method because of its potential to facilitate the disruption of various lignocellulosic materials. Use of ultrasound can scale down the hydrolysis time of biomass up to 80% providing advantage to bio-fuel production (Luo et al. 2014). But, the process is energy intensive and systematic studies are needed for optimizing the process parameters for commercial applications.

References

Agbor VB, Cicek N, Sparling R, Berlin A, Levin DB (2011) Biomass pretreatment: fundamentals toward application. Biotechnol Adv 29:675–685

Alizadeh H, Teymouri F, Gilbert TI, Dale BE (2005) Pretreatment of switchgrass by ammonia fibre explosion (AFEX). Appl Biochem Biotechnol 121–124:1133–1141

Alriols MG, Tejado A, Blanco M, Mondragon I, Labidi J (2009) Agricultural palmoil tree residues as raw material for cellulose, lignin and hemicelluloses production by ethylene glycol pulping process. Chem Eng J 148(1):106–114

Alriols MG, Garcıa A, Llano-ponte R, Labidi J (2010) Combined organosolv and ultrafiltration lignocellulosic biorefinery process. Chem Eng J 157(1):113–120

Alvira P, Tomás-Pejó E, Ballesteros M, Negro MJ (2010) Pretreatment technologies for an efficient bioethanol production process based on enzymatic hydrolysis: a review. Bioresour Technol 101(13):4851–4861

Antal MJ Jr (1996) Water: a traditional solvent pregnant with new application. In: White HJ Jr (ed) Proceedings of the 12th international conference on the properties of water and steam. Begell House, New York, pp 23–32

Arato C, Pye EK, Gjennestad G (2005) The lignol approach to biorefining of woody biomass to produce ethanol and chemicals. Appl Biochem Biotechnol A 123(1–3):871–882

Avella R, Scoditti E (1998) The Italian steam explosion program of ENEA. Biomass Bioenergy 14:289–293

Bajpai P (2016) In pretreatment of lignocellulosic biomass for biofuel production. Springer Briefs in Molecular Science, Springer Nature

Bak JS, Ko JK, Han YH (2009) Improved enzymatic hydrolysis yield of rice straw using electron beam irradiation pretreatment. Bioresour Technol 100:1285–1290

Banerjee G, Car S, Liu T (2012) Scale-up and integration of alkaline hydrogen peroxide pretreatment, enzymatic hydrolysis, and ethanolic fermentation. Biotechnol Bioeng 109(4):922–931

Baruah J, Nath BK, Sharma R, Kumar S, Deka RC, Baruah DC, Kalita E (2018) Recent trends in the pretreatment of lignocellulosic biomass for value-added products. Front Energy Res 6.141

Ben-Ghedalia D, Miron J (1981) The effect of combined chemical and enzyme treatment on the saccharification and in vitro digestion rate of wheat straw. Biotechnol Bioeng 1981(23):823–831

Bensah EC, Mensah M (2013a) Chemical pretreatment methods for the production of cellulosic ethanol: technologies and innovations. Int J Chem Eng 2013:21, Article ID719607

Bensah EC, Mensah M (2013b) Chemical pretreatment methods for the production of cellulosic ethanol: technologies and innovations. Int J Chem Eng 2013:21

Bjerre AB, Olesen AB, Fernqvist T (1996) Pretreatment of wheat straw using combined wet oxidation and alkaline hydrolysis resulting in convertible cellulose and hemicellulose. Biotechnol Bioeng 49:568–577

Brosse N, Hage REL, Sannigrahi P, Ragauskas A (2010) Dilute sulphuric acid and ethanol organosolv pretreatment of miscanthus x giganteus. Cellul Chem Technol 44(1–3):71–78

Cadoche L, López GD (1989) Assesment of size reduction as a preliminary step in the production of ethanol from lignocellulosic wastes. Biotechnol Wastes 30:153–7

Cai J, Zhang L, Zhou J, Li H, Chen H, Jin H (2004) Novel fibers prepared from cellulose in NaOH/urea aqueous solution. Macromol Rapid Commun 25:1558–1562

Cao Y, Li H, Zhang Y, Zhang J, He J (2010) Structure and properties of novel regenerated cellulose films prepared from cornhusk cellulose in room temperature ionic liquids. J Appl Polymer Sci 116:547–554

Chandra RP, Bura R, Mabee WE (2007) Substrate pretreatment: the key to effective enzymatic hydrolysis of lignocellulosics? Adv Biochem Eng/Biotechnol 108:67–93

Chang VS, Holtzapple MT (2000) Fundamental factors affecting biomass enzymatic reactivity. Appl Biochem Biotechnol 84–86:5–37

Chang MM, Chou TYC, Tsao GT (1981) Structure, pretreatment and hydrolysis of cellulose. Adv Biochem Eng 20:15–42

Chang VS, Burr B, Holtzapple MT (1997) Lime pretreatment of switchgrass. Appl Biochem Biotechnol 63–65:3–19

Chan YL, Yang J, Ahn JW, Moon YH, Yoon YM, Gyeong-Dan Yu GD, An GH, Choi IH (2014) The optimized CO_2-added ammonia explosion pretreatment for bioethanol production from rice straw. Bioprocess Biosyst Eng 37(9):1907–1915

Chaturvedi V, Verma P (2013) An overview of key pretreatment processes employed for bioconversion of lignocellulosic biomass into biofuels and value added products. Biotech 3:415–431

Cheah WY, Sankaran R, Show PL, Tg Ibrahim TNB, Chew KW, Culaba A, Chang J-S (2020) Pretreatment methods for lignocellulosic biofuels production: current advances, challenges and future prospects. Biofuel Res J 25:1115–1127. https://doi.org/10.18331/brj2020.7.1.4

Cherpozat L, Loranger E, Daneault C (2017) Ultrasonic pretreatment effects on the bio-oil yield of a laboratory-scale slow wood pyrolysis. J Anal Appl Pyrolysis 126:31–38. https://doi.org/10.1016/j.jaap.2017.06.027

Chum HL, Douglas LJ, Feinberg DA, Schroeder HA (1985) Evaluation of pretreatments of biomass for enzymatic hydrolysis of cellulose. US Department of Energy, ContractNo.DE-AC02-83CHt0093

Conde-Mej´ıa C, Jim´enez-Guti´erreza A, El-Halwagi M (2012) A comparison of pretreatment methods for bioethanol production from lignocellulosic materials. Saf Environ Prot 90(3):189–202

Conte P, Maccotta A, Pasquale CD, Bubici S, Alonzo G (2009) Dissolution mechanism of crystalline cellulose in H3PO4 as assessed by high field NMR spectroscopy and fast field cycling NMR relaxometry. J Agric Food Chem 57:8748–8752

Dale BE, Moreira MJ (1982) A freeze-explosion technique for increasing cellulose hydrolysis. Biotechnol Bioeng Symp 12:31–43

Dale BE, Henk LL, Shiang M (1984) Fermentation of lignocellulose materials treated my ammonia freeze-explosion. Dev Ind Microbiol 26:223–233

Damaso MCT, Mde C, Castro RM, Andrade MC, Pereira N (2004) Application of xylanase from Thermomyces lanuginosus IOC-4145 for enzymatic hydrolysis of corn cob and sugarcane bagasse. Appl Biochem Biotechnol 115:1003–1012

Delgenes JP, Penaud V, Moletta R (2002) Pretreatment for the enhancement of anaerobic digestion of solid waster Chap. 8. In: Biomethanization of the organic fraction of municipal solid waste. IWA publishing, pp 201–228

Duff SJB, Murray WD (1996) Bioconversion of forest products industry waste cellulosics to fuel ethanol: a review. Bioresour Technol 55:1–33

Dunlap CE, Chiang LC (1980) Cellulose degradation-a common link. In: Shuler ML (ed) Utilization and recycle of agricultural wastes and residues. CRC Press, Boca Raton, FL, pp 19–65

Eggeman T, Elander TR (2005) Process and economic analysis of pretreatment technologies. Biores Technol 8:2019–2025

Esteghalian Torget RW, Himmel M, Grohmann K (1992) Dilute acid pretreatment of two short-rotation herbaceous crops. Appl Biochem Biotechnol 34(35):115–23

Fan LT, Gharpuray MM, Lee YH (1987) Cellulose hydrolysis biotechnology monographs. Springer, Berlin, pp 57

Fink HP, Weigel P, Purz HJ (2001) Structure formation of regenerated cellulose materials from NMMO-solutions. J Progr Polym Sci 26:1473–1524
Foster BL, Dale BE, Peterson JBD (2001) Enzymatic hydrolysis of ammonia-treated sugar beet pulp. Appl Biochem Biotechnol 91–93:269–282
Galbe M, Zacchi G (2007) Pretreatment of lignocellulosic materials for efficient bioethanol production. Biofuels 108:41–65
Ge X, Green VS, Zhang N, Sivakumar G, Xu J (2012) Eastern gamagrass as an alternative cellulosic feedstock for bioethanol production. Process Biochem 47:335–339
Gregg DJ, Saddler JN (1996) Factors affecting cellulose hydrolysis and the potential of enzyme recycle to enhance the efficiency of an integrated wood to ethanol process. Biotechnol Bioeng 51:375–383
Grous WR, Converse AO, Grethlein HE (1986) Effect of steam explosion pretreatment on pore size and enzymatic hydrolysis of poplar. Enzym Microb Technol 8:274–280
Guragain YN, De Coninck J, Husson F, Durand A, Rakshit SK (2011) Comparison of some new pretreatment methods for second generation bioethanol production from wheat straw and water hyacinth. Bioresour Technol 102(6):4416–4424
Hammel KE, Kapich AN, Jensen KA Jr, Ryan ZC (2002) Reactive oxygen species as agents of wood decay by fungi. Enzym Microb Technol 30(4):445–453
Hatakka A (1994) Lignin-modifying enzymes from selected white-rot fungi: production and role from in lignin degradation. FEMS Microbiol Rev 13:125–35
Hatakka AI, Varesa T, Lunn TK (1993) Production of multiple lignin peroxidases by the whiterot fungus Phlebia ochraceofulva. Enzyme Microb Technol 15:664–9
Haverty D, Dussan K, Piterina AV, Leahy JJ, Hayes MHB (2012) Autothermal, single-stage, performic acid pretreatment of Miscanthus x giganteus for the rapid fractionation of its biomass components into a lignin/hemicellulose-rich liquor and a cellulase-digestible pulp. Bioresour Technol 109:173–177
Heinze T, Schwikal K, Barthel S (2005) Ionic liquids as reaction medium in cellulose functionalization. Macromol Biosci 5:520–525
Hendricks AT, Zeeman G (2009) Pretreatments to enhance the digestibility of lignocellulosic biomass. Bioresour Technol 100:10–18
He Z, Wang Z, Zhao Z, Yi S, Mu J, Wang X (2017) Influence of ultrasound pretreatment on wood physiochemical structure. Ultrason Sonochem 34:136–141. https://doi.org/10.1016/j.ultsonch.2016.05.035
Hideno A, Kawashima A, Endo T, Honda K, Morita M (2013) Ethanol-based organosolv treatment with trace hydrochloric acid improves the enzymatic digestibility of Japanese cypress (Chamaecyparis obtusa). by exposing nanofibers on the surface. Bioresour Technol 18:64–70
Hinman ND, Schell DJ, Riley CJ, Bergeron PW, Walter PJ (1992) Preliminary estimate of the cost of ethanol production for SSF technology. Appl Biochem Biotechnol 34(35):639–649
Holtzapple MT, Davison RR, Stuart ED (1992b) Biomass refining process. US Patent 5(171):592
Holtzapple MT, Jun JH, Ashok G, Patibandla SL, Dale BE (1990) Ammonia fiber explosion (AFEX) pretreatment of lignocellulosic wastes. In: American Institute of Chemical Engineers National Meeting. Chicago, IL
Holtzapple MT, Jun JH, Ashok G, Patibandla SL, Dale BE (1991) The ammonia freeze explosion (AFEX) process: a practical lignocellulose pretreatment. Appl Biochem Biotechnol 28–29:59–74
Holtzapple MT, Lundeen JE, Sturgis R, Lewis JE, Dale BE (1992a) Pretreatment of lignocellulosic municipal solid-waste by ammonia fibre explosion (AFEX). Appl Biochem Biotechnol 34–35:5–21
Huijgen WJJ, Van der Laan RR, Reith JH (2008). Mordified organosolv as a fractionation process of lignocellulosic biomass for coproduction of fuels and chemicals. In: Proceedings of the 16th Europeran biomass conference and exhibition. Valencia, Spain
Hwang SS, Lee SJ, Kim HK (2008) Biodegradation and saccharification of wood chips of Pinus strobus and Liriodendron tulipifera by white rot fungi. J Microbiol Biotechnol 18:1819–1825

Ichwan M, Son TW (2011) Study on organosolv pulping methods of oil palm biomass. In: International seminar on chemistr. pp 364–370
Imai M, Ikari K, Suzuki I (2004) High-performance hydrolysis of cellulose using mixed cellulase species and ultrasonication pretreatment. Biochem Eng J 17:79–83
Isci A, Himmelsbach JN, Pometto AL, Raman R, Anex RP (2008) Aqueous ammonia soaking of switchgrass followed by simultaneous saccharification and fermentation. Appl Biochem Biotechnol 144:69–77
Itoh H, Wada M, Honda Y, Kuwahara M (2003) Bioorganosolve pretreatments for simultaneous saccharification and fermentation of beech wood by ethanolysis and white-rot fungi. J Biotechnol 103 (3):273–280
Ivetić DŽ, Omorjan RP, Ordevi, ć TR, Antov MG (2017) The impact of ultrasound pretreatment on the enzymatic hydrolysis of cellulose from sugar beet shreds: modeling of the experimental results. Environ Prog Sustain Energy 36:1164–1172. https://doi.org/10.1002/ep.12544
Iyer PV, Wu ZW, Kim SB, Lee YY (1996) Ammonia recycled percolation process for pretreatment of herbaceous biomass. Appl Biochem Biotechnol 57–58:121–132
Jorgensen H, Kristensen JB, Felby C (2007) Enzymatic conversion of lignocellulose into fermentable sugars: challenges and opportunities. Biofuels Bioprod Biorefining 1(2):119-134
Kappe CO (2004) Controlled microwave heating in modern organic synthesis. Angew Chem Int Ed 43:6250–6284
Keller FA, Hamilton JE, Nguyen QA (2003) Microbial pretreatment of biomass-potential for reducing severity of thermochemical biomass pretreatment. Appl Biochem Biotechnol 105:27–41
Khodaverdi M, Jeihanipour A, Karimi K, Taherzadeh MJ (2012) Kinetic modeling of rapid enzymatic hydrolysis of crystalline cellulose after pretreatment by NMMO. J Ind Microbiol Biotechnol 39:429–438
Kilzer FJ, Broido A (1965) Speculations on the nature of cellulose pyrolysis. Pyrodynamics 2:151–163
Kim S, Holtzapple MT (2006a) Effect of structural features on enzyme digestibility of corn stover. Bioresour Technol 97:583–591
Kim S, Holtzapple MT (2006b) Lime pretreatment and enzymatic hydrolysis of corn stover. Bioresour Technol 96:1994–2006
Kim HK, Hong J (2001) Supercritical CO_2 pretreatment of lignocellulose enhances enzymatic cellulose hydrolysis. Bioresour Technol 77:139–144
Kim SB, Lee YY (2002) Diffusion of sulfuric acid within lignocellulosic biomass particles and its impact on dilute-acid pretreatment. Bioresour Technol 83:165–171
Kim TH, Lee YY (2005a) Pretreatment and factionation of corn stover by soaking in aqueous ammonia. Appl Biochem Biotechnol 121:1119–31
Kim TH, Lee YY (2005b) Pretreatment of corn stover by ammonia recycle percolation process. Bioresour Technol 96:2007–13
Kim SB, Yum DM, Park SC (2000) Step-change variation of acid concentration in a percolationreactor for hydrolysis of hardwood hemicellulose. Bioresour Technol 72:289–94
Kim HT, Kim JS, Sunwoo C, Lee YY (2003) Pretreatment of corn stover by aqueous ammonia. Bioresour Technol 90:39–47
Kim TH, Lee YL, Sunwoo C, Kim JS (2006) Pretreatment of corn stover by low-liquid ammonia recycle percolation process. Appl Biochem Biotechnol 133:41–57
Kim Y, Yu A, Han M, Choi GW, Chung B (2011) Enhanced enzymatic saccharification of barley strawpretreated by ethanosolv technology. Appl Biochem Biotechnol 163(1):143–152
Kim HY, Gwak KS, Lee SY, Jeong HS, Ryu KO, Choi IG (2012) Biomass characteristics and ethanol production of yellow poplar (Liriodendron tulipifera) treated with slurry composting and biofiltration liquid as fertilizer. Biomass Bioenergy 42:10–17
Kitchaiya P, Intanakul P, Krairiksh MJ (2003) Enhancement of enzymatic hydrolysis of lignocellulosic wastes by microwave pretreatment under atmospheric-pressure. J Wood Chem Technol 23:217–225

Kohlmann KL, Sariyaka A, Westgate PJ, Weil J, Velayudhan A, Hendrickson R, Ladisch MR (1995) Enhance enzyme activities on hydrated lignocellulosic substrates. In: Saddler JN, Penner MH (eds) Enzymatic degradation of insoluble carbohydrates. ACS publishing, pp 237–255

Koo BW, Min BC, Gwak KS (2012) Structural changes in lignin during organosolv pretreatment of liriodendron tulipifera and the effect on enzymatic hydrolysis. Biomass Bioenergy 42:24–32

Koutsianitis D, Mitani C, Giagli K, Tsalagkas D, Halász K, Kolonics O et al (2015) Properties of ultrasound extracted bicomponent lignocellulose thin films. Ultrason Sonochem 23:148–155. https://doi.org/10.1016/j.ultsonch.2014.10.014

Kuhad RC, Singh A, Ericksson KE (1997) Microorganisms and enzymes involved in the degradation of plant fiber cell walls. Adv Biochem Eng Biotechnol 57:45–125

Kuhar S, Nair LM, Kuhad RC (2008) Pretreatment of lignocellulosic material with fungi capable of higher lignin degradation and lower carbohydrate degradation improves substrate acid hydrolysis and the eventual conversion to ethanol. Can J Microbiol 54:305–313

Kumar P, Barrett DM, Delwiche MJ, Stroeve P (2009) Methods for pretreatment of lignocellulosic biomass for efficient hydrolysis and biofuel production. Ind Eng Chem Res 48:3713–3729

Kuo C-H, Lee C-K (2009) Enhancement of enzymatic saccharification of cellulose by cellulose dissolution pretreatments. Carbohydr Polym 77:41–46

Kupiainen L, Ahola J, Tanskanen J (2012) Hydrolysis of organosolv wheat pulp in formic acid at high temperature for glucose production. Bioresour Technol 116:29–35

Ladisch MR, Ladisch CM, Tsao GT (1978) Cellulose to sugars: new path gives quantitative yield. Science 201:743–745

Lee J (1997) Biological conversion of lignocellulosic biomass to ethanol. J Biotechnol 56:1–24

Lee JW, Gwak KS, Park JY (2007) Biological pretreatment of softwood Pinus densiflora by three white rot fungi. J Microbiol 45:485–491

Lin KW, Ladisch MR, Schaefer DM (1981) Review on effect of pretreatment on digestibility of cellulosic materials. AIChE Symp 207:102–106

Liu J, Takada R, Karita S, Watanabe T, Honda Y, Watanabe T (2010) Microwave-assisted pretreatment of recalcitrant softwood in aqueous glycerol. Bioresour Technol 23:9355–9360

Liyakathali NAM, Muley PD, Aita G, Boldor D (2016) Effect of frequency and reaction time in focused ultrasonic pretreatment of energy cane bagasse for bioethanol production. Bioresour Technol 200:262–271. https://doi.org/10.1016/j.biortech.2015.10.028

Li F, Yao R, Wang H, Hu H, Zhang R (2012a) Process optimization for sugars production from rice straw via pretreatment with sulphur trioxide micro-thermal explosion. BioResources 7(3):3355–3366

Li Q, Ji GS, Tang YB, Gu XD, Fei JJ, Jiang HQ (2012b) Ultrasound-assisted compatible in situ hydrolysis of sugarcane bagasse in cellulase-aqueous-N-methylmorpholine- N-oxide system for improved saccharification. Bioresour Technol 107:251–257

Lucas M, Hanson SK, Wagner GL, Kimball DB, Rector KD (2012) Evidence for room temperature delignification of wood using hydrogen peroxide and manganese acetate as a catalyst. Bioresour Technol 119:174–180

Luo J, Fang Z, Smith RL Jr (2014) Ultrasound-enhanced conversion of biomass to biofuels. Prog Energy Combust Sci 41:56–93. https://doi.org/10.1016/j.pecs.2013.11.001

Luzzi SC, Artifon W, Piovesan B, Tozetto E, Mulinari J, Kuhn GDO et al (2017) Pretreatment of lignocellulosic biomass using ultrasound aiming at obtaining fermentable sugar. Biocatal Biotransform 35:161–167. https://doi.org/10.1080/10242422.2017.1310206

Maa H, Liu W, Chen X, Wua Y, Yu Z (2009) Enhanced enzymatic saccharification of rice straw by microwave pretreatment. Bioresour Technol 100:1279–1284

McMillan JD (1994a) Pretreatment of lignocellulosic biomass. Enzym Convers Biomass Fuels Prod 566:292–324

McMillan JD (1994b) Pretreatment of lignocellulosic biomass. In: Himmel ME, Baker JO, Overend RP (eds) Enzymatic conversion of biomass for fuels production. American Chemical Society, Washington, D.C., pp 292–324

Mesa L, Gonz´alez E, Cara C, Gonz´alez M, Castro E, Mussatto SI (2011) The effect of organosolv pretreatment variables on enzymatic hydrolysis of sugarcane bagasse. Chem Eng J 168(3):1157–1162
Mes-Hartree M, Dale BE, Craig WK (1988) Comparison of steam and ammonia pretreatment for enzymatic hydrolysis of cellulose. Appl Microbiol Biotechnol 29:462–468
Millet MA, Baker AJ, Satter LD (1976) Physical and chemical pretreatments for enhancing cellulose saccharification. Biotech Bioeng Symp 6:125–153
Mok WSL, Antal MJ Jr (1992) Uncatalyzed solvolysis of whole biomass hemicellulose by hot compresses liquid water. Ind Eng Chem Res 31:1157–1161
Mok WSL, Antal MJ Jr (1994) Biomass fractionation by hot compressed liquid water. In: Bridgewater AV (ed) Advances in thermochemical biomass conversion, vol 2. Blackie Academic and Professional Publishers, New York, pp 1572–1582
Mosier N, Hendrickson R, Brewer M, Ho N, Sedlak M, Dreshel R (2005a) Industrial scale-up of pH-controlled liquid hot water pretreatment of corn fiber for fuel ethanol production. Appl Biochem Biotechnol 125:77–97
Mosier N, Wyman CE, Dale BE, Elander R, Lee YY, Holtzapple MT (2005b) Features of promising technologies for pretreatment of lignocellulosic biomass. Bioresour Technol 96:673–86
Moxley GM, Zhu Z, Zhang Y-HP (2008) Efficient sugar release by the cellulose solvent based lignocellulose fractionation technology and enzymatic cellulose hydrolysis. J Agric Food Chem 56:7885–7890
Murnen HK, Balan V, Chundawat SPS, Bals B, Sousa LDC, Dale BE (2007) Optimization of ammonia fiber expansion (AFEX) pretreatment and enzymatic hydrolysis of miscanthus x giganteus to fermentable sugars. Biotechnol Prog 23:846–850
Nakamura Y, Daidai M, Kobayashi F (2004) Ozonolysis mechanism of lignin model compounds and microbial treatment of organic acids produced. Water Sci Technol 50(3):167–172
Nakashima K, Ebi Y, Shibasaki-Kitakawa N, Soyama H, Yonemoto T (2016) Hydrodynamic cavitation reactor for efficient pretreatment of lignocellulosic biomass. Ind Eng Chem Res 55:1866–1871. https://doi.org/10.1021/acs.iecr.5b04375
Narayanaswamy N, Faik A, Goetz DJ, Gu T (2011) Supercritical carbon dioxide pretreatment of corn stover and switchgrass for lignocellulosic ethanol production. Bioresour Technol 102(2011):6995–7000
Neely WC (1984) Factors affecting the pretreatment of biomass with gaseous ozone. Biotechnol Bioeng 26:59–65
Nguyen QA, Tucker MP, Keller FA, Eddy FP (2000) Two-stage dilute-acid pretreatment of softwoods. Appl Biochem Biotechnol 84–86:561–575
Nitayavardhana S, Rakshit SK, Grewell D (2008) Ultrasound pretreatment of cassava chip slurry to enhance sugar release for subsequent ethanol production. Biotechnol Bioeng 101:487–496
Palmowski L, Muller J (1999) Influence of the size reduction of organic waste on their anaerobic digestion. In: II international symposium on anaerobic digestion of solid waste. Barcelona 15–17 June, pp 137–144
Palmqvist E, Hahn-Hägerdal B (2000a) Fermentation of lignocellulosic hydrolysates I: inhibition and detoxification. Bioresour Technol 74:17–24
Pan X, Arato C, Gilkes N (2005) Biorefining of softwoods using ethanol organosolv pulping: preliminary evaluation of process streams for manufacture of fuel-grade ethanol and coproducts. Biotechnol Bioeng 90(4):473–481
Pan X, Gilkes N, Kadla J (2006a) Bioconversion of hybrid poplar to ethanol and co-products using an organosolv fractionation process: optimization of process yields. Biotechnol Bioeng 94(5):851–861
Pan X, Kadla JF, Ehara K, Gilkes N, Saddler JN (2006b) Organosolv ethanol lignin from hybrid poplar as a radical scavenger: relationship between lignin structure, extraction conditions, and antioxidant activity. J Agric Food Chem 54(16):5806–5813

Park N, Kim HY, Koo BW, Yeo H, Choi IG (2010) Organosolv pretreatment with various catalysts for enhancing enzymatic hydrolysis of pitch pine (Pinus rigida). Bioresour Technol 101(18):7046–7053

Prado R, Erdocia X, Serrano L, Labidi J (2012) Lignin purification with green solvents. Cellul Chem Technol 46(3–4):221–225

Pu YQ, Jiang N, Ragauskas AJ (2007) Ionic liquids as a green solvent for lignin. J Wood Chem Technol 27:23–33

Pye EK, Lora JH (1991) The AlcellTM process: a proven alternative to kraft pulping, Tappi J 113

Ravindran R, Jaiswal AK (2016) A comprehensive review on pretreatment strategy for lignocellulosic food industry waste: challenges and opportunities. Bioresour Technol 199:92–102. https://doi.org/10.1016/j.biortech.2015.07.106

Reshamwala S, Shawky BT, Dale BE (1995) Ethanol production from enzymatic hydrolysates of AFEX-treated coastal Bermuda grass and switchgrass. Appl Biochem Biotechnol 51(52):43–55

Roberts BA, Strauss CR (2005) Toward rapid, green, predictable microwave-assisted synthesis. Acc Chem Res 38:653–661

Rollin JA, Zhu Z, Sathisuksanoh N, Zhang Y-HP (2011) Increasing cellulose accessibility is more important than removing lignin: a comparison of cellulose solvent-based lignocellulose fractionation and soaking in aqueous ammonia. Biotechnol Bioeng 108:22–30

Ruan D, Zhang L, Zhou J, Jin H, Chen H (2004) Structure and properties of novel fibers spun from cellulose in NaOH/thiourea aqueous solution. Macromol Biosci 4:1105–1112

Ryu DDY, Lee SB (1982) Effect of compression milling on cellulose structure and on enzymatic hydrolysis kinetics. Biotechnol Bioeng 24:1047–1067

Saddler JN, Ramos LP, Breuil C (1993) Steam pretreatment of lignocellulosic residues (Chap. 3). In: Saddler JN (ed) Bioconversion of forest and agricultural plant residues. CAB International, Oxford, UK, pp 73—92

Saha BC, Biswas A, Cotta MA (2008) Microwave pretreatment enzymatic saccharification and fermentation of wheat straw to ethanol. J Biobased Mater Bioenergy 2:210–217

Sarkar N, Ghosh SK, Bannerjee S, Aikat K (2012) Bioethanol production from agricultural wastes: an overview. Renew Energy 37(1):19–27

Sathitsuksanoh N, Zhu Z, Templeton N, Rollin J, Harvey S, Zhang Y-HP (2009) Saccharification of a potential bioenergy crop, Phragmites australis (common reed), by lignocellulose fractionation followed by enzymatic hydrolysis at decreased cellulase loadings. Ind Eng Chem Res 48:6441–6447

Sathitsuksanoh N, Zhu Z, Ho T-J, Bai M-D, Zhang Y-HP (2010) Bamboo saccharification through cellulose solvent-based biomass pretreatment followed by enzymatic hydrolysis at ultra-low cellulase loadings. Bioresource Technol 101:4926–4929

Sathitsuksanoh N, Zhu ZG, Wi S, Zhang Y-HP (2011) Cellulose solvent based biomass pretreatment breaks highly ordered hydrogen bonds in cellulose fibers of switchgrass. Biotechnol Bioeng 108:521–529

Sathitsuksanoh N, Zhu Z, Zhang Y-HP (2012a) Cellulose solvent and organic solvent-based lignocellulose fractionation enabled efficient sugar release from a variety of lignocellulosic feedstocks. Bioresource Technol 117:228–233

Sathitsuksanoh N, George A, Zhang Y-HP (2012b) New lignocellulose pretreatments using cellulose solvents: a review. J Chem Technol Biotechnol 88:169–180

Shafiei M, Karimi K, Taherzadeh MJ (2011) Technoeconomical study of ethanol and biogas from spruce wood by NMMO-pretreatment and rapid fermentation and digestion. Bioresour Technol 102(17):7879–7886

Shafizadeh F, Bradbury AGW (1979) Thermal degradation of cellulose in air and nitrogen at low temperatures. J Appl Polym Sci 23:1431–1442

Shafizadeh F, Lai YZ (1975) Thermal degradation of 2-deoxy-Darabino-hexonic acid and 3-deoxy-D-ribo-hexono-1,4-lactone. Carbohydr Res 42:39–53

Sheehan J, Aden A, Paustian K, Killian K, Brenner J, Walsh M, Nelson R (2003) Energy and environmental aspects of using corn stover for fuel ethanol. J Ind Ecol 7:117–146

Shin SJ, Sung YJ (2008) Improving enzymatic hydrolysis of industrial hemp (Cannabis sativa L.) by electron beam irradiation. Radiat Phys Chem 77:1034–1038

Shi J, Chinn MS, Sharma-Shivappa RR (2008) Microbial pretreatment of cotton stalks by solid state cultivation of Phanerochaete chrysosporium. Bioresour Technol 99:6556–6564

Singh P, Suman A, Tiwari P (2008) Biological pretreatment of sugarcane trash for its conversion to fermentable sugars. World J Microbiol Biotechnol 24:667–673

Sivers MV, Zacchi G (1995) A techno-economical comparison of three processes for the production of ethanol from pine. Bioresour Technol 51:43–52

Socha AM, Parthasarathi R, Shi J, Pattathil S, Whyte D, Bergeron M, George A, Tran K, Stavila V, Venkatachalam S, Hahn MG, Simmons BA, Singh S (2014) Efficient biomass pretreatment using ionic liquids derived from lignin and hemicellulose. PNAS. https://doi.org/10.1073/pnas.1405685111

Striegel AM (1997) Theory and applications of DMAc/LiCl in the analysis of polysacharrides. Carbohydr Polym 34:267–274

Subhedar PB, Gogate PR (2014) Alkaline and ultrasound assisted alkaline pretreatment for intensification of delignification process from sustainable raw-material. Ultrason Sonochem 21:216–225. https://doi.org/10.1016/j.ultsonch.2013.08.001

Sun Y, Cheng J (2002) Hydrolysis of lignocellulosic materials for ethanol production: a review. Bioresour Technol 83:1–11

Su MY, Tzeng WS, Shyu YT (2010) An analysis of feasibility of bioethanol production from Taiwan sorghum liquor waste. Bioresour Technol 101(17):6669–6675

Szijártó N, Kádár Z, Varga E, Thomsen AB, Costa-Ferreira M, Re´czey K (2009) Pretreatment of reed by wet oxidation and subsequent utilization of the pretreated fibers for ethanol production. Appl Biochem Biotechnol 155(1–3):386–396

Tengerdy RP, Nagy JG (1988) Increasing the feed value of forestry waste by ammonia freeze explosion treatment. Biol Wastes 25:149–153

Torget RW, Himmel M, Grohmann K (1992) Dilute acid pretreatment of two short-rotation herbaceous crops. Appl Biochem Biotechnol 34(35):115–123

Tucker MP, Kim KH, Newman MM, Nguyen QA (2003) Effects of temperature and moisture on dilute-acid steam explosion pretreatment of corn stover and cellulase enzyme digestibility. Appl Biochem Biotechnol 10:105–108

Ungurean M, Fitigau F, Paul C, Ursoiu A, Peter F (2011) Ionic liquid pretreatment and enzymatic hydrolysis of wood biomass. World Acad Sci Eng Technol 76:387–391

Uppal SK, Kaur R, Sharma P (2011) Optimization of chemical pretreatment and acid saccharification for conversion of sugarcane bagasse to ethanol. Sugar Tech 13(3):214–219

Varga E, Schmidt AS, Re´czey K, Thomsen AB (2003) Pretreatment of corn stover using wet oxidation to enhance enzymatic digestibility. Appl Biochem Biotechnol 104(1):37–50

Vidal PF, Molinier J (1988) Ozonolysis of lignins Improvement of in vitro digestibility of poplar sawdust. Biomass 1988(16):1–17

Villaverde JJ, Ligero P, de Vega A (2010) Miscanthus x giganteus as a source of biobased products through organosolv fractionation: a mini review. Open Agric J 4:102–110

Vlasenko EY, Ding H, Labavitch JM, Shoemaker SP (1997) Enzymatic hydrolysis of pretreated rice straw. Bioresour Technol 59:109–119

Wang Y, Zhao Y, Deng Y (2008) Effect of enzymatic treatment on cotton fiber dissolution in NaOH/urea solution at cold temperature. Carbohydr Polym 72:178–184

Wang K, Yang HY, Xu F, Sun RC (2011a) Structural comparison and enhanced enzymatic hydrolysis of the cellulosic preparation from Populus tomentosa Carr. by different cellulose-soluble solvent systems. Bioresource Technol 102:4524–4529

Wang C, Zhou F, Yang Z (2012) Hydrolysis of cellulose into reducing sugar via hot-compressed ethanol/water mixture. Biomass Bioenerg 42:143–150

Weil JR, Sariyaka A, Rau SL, Goetz J, Ladisch CM, Brewer M (1997) Pretreatment of yellow poplar wood sawdust by pressure cooking in water. Appl Biochem Biotechnol 68:21–40

Weil J, Brewer M, Hendrickson R, Sariyaka A, Ladisch MR (1998) Continuous pH monitoring-during pretreatment of yellow poplar wood sawdust by pressure cooking in water. Appl Biochem Biotechnol 70–72:99–111

Wilson NS, Sarko CR, Roth GP (2004) Development and applications of a practical continuous flow. Microwave cell. Org Proc Res Dev 8(3):535–538.24

Wojciak A, Pekarovicova A (2001) Enhancement of softwood kraft pulp accessibility for enzymatic hydrolysis by means of ultrasonic irradiation. Cellul Chem Technol 35:361–369

Wright JD (1998) Ethanol from biomass by enzymatic hydrolysis. Chem Eng Prog 84(8):62–74

Wu J, Zhang J, He J (2004) Homogeneous acetylation of cellulose in a new ionic liquid. Biomacromol 5:266–268

Wyman CE, Dale BE, Elander RT, Holtzapple M, Ladisch MR, Lee YY (2005a) Comparative sugar recovery data from laboratory scale application of leading pretreatment technologies to corn stover. Bioresour Technol 96(18):2026–2032

Wyman CE, Dale BE, Elander RT, Holtzapple MT, Ladisch MR, Lee YY (2005b) Coordinated development of leading biomass pretreatment technologies. Bioresour Technol 96:1959–66

Yang B, Wyman CE (2004) Effect of xylan and lignin removal by batch and flow through pretreatment on enzymatic digestibility of corn stover cellulose. Biotechnol Bioeng 86:88–95

Yang B, Wyman CE (2008) Pretreatment: the key to unlocking low-cost cellulosic ethanol. Biofuels, Bioprod Biorefin 2(1):26–40

Yang F, Hanna M, Sun R (2012) Value-added uses for crude glycerol—a byproduct of biodiesel production. Biotechnol Biofuels 5(13)

Yao RS, Hu HJ, Deng SS, Wang H, Zhu HX (2011) Structure and saccharification of rice straw pretreated with sulfur trioxide micro-thermal explosion collaborative dilutes alkali. Bioresour Technol 102(10):6340–6343

Youssef BM, Aziz NH (1999) Influence of gamma-irradiation on the bioconversion of rice straw by Trichoderma viride into single cell protein. Cytobios 97:171–183

Yu J, Zhang J, He J, Liu Z, Yu Z (2009) Combinations of mild physical or chemical pretreatment with biological pretreatment for enzymatic hydrolysis of rice hull. Bioresour Technol 100:903–908

Zhang Y-HP (2008) Reviving the carbohydrate economy via multi-product biorefineries. J Ind Microbiol Biotechnol 35:367–375

Zhang Y-HP, Lynd LR (2006) A functionally based model for hydrolysis of cellulose by fungal cellulase. Biotechnol Bioeng 94:888–98

Zhang Y-HP, Cui J, Lynd LR, Kuang LR (2006) A transition from cellulose swelling to cellulose dissolution by o-phosphoric acid: evidence from enzymatic hydrolysis and supramolecular structure. Biomacromol 7:644–648

Zhang XY, Xu CY, Wang HXJ (2007a) Pretreatment of bamboo residues with Coriolus versicolor for enzymatic hydrolysis. J Biosci Bioeng 104:149–151

Zhang XY, Yu HB, Huang HY (2007b) Evaluation of biological pretreatment with white rot fungi for the enzymatic hydrolysis of bamboo culms. Int Biodeterior Biodegrad 60:159–164

Zhang S, Marechal F, Gassner M, Perin-Levasseur Z, Qi W, Ren Z, Yan Y, Farvat D (2009a) process modelling and integration of fuel ethanol production from lignocellulosic biomass based on double acid hydrolysis. Energy Fuels. https://doi.org/10.1021/ef801027x

Zhang S, Marechal F, Gassner M, Perin-Levasseur Z, Qi W, Ren Z, Yan Y, Farvat D (2009b) process modelling and integration of fuel ethanol production from lignocellulosic biomass based on double acid hydrolysis. Energy Fuels. https://doi.org/10.1021/ef801027x

Zhang Z, O'Hara IM, Doherty WOS (2012) Pretreatment of sugarcane bagasse by acid-catalysed process in aqueous ionic liquid solutions. Bioresour Technol 120:149–156

Zhang T, Kumar R, Wyman CE (2013) Sugar yields from dilute oxalic acid pretreatment of maple wood compared to those with other dilute acids and hot water. Carbohydr Polym 92:334–344

Zhao X, Liu D (2012) Fractionating pretreatment of sugarcane bagasse by aqueous formic acid with direct recycle of spent liquor to increase cellulose digestibility-the Formiline process. Bioresour Technol 117:25–32

Zhao H, Kwak JH, Wang Y, Franz JA, White JM, Holladay JE (2007) Interactions between cellulose and N-methylmorpholine- N-oxide. Carbohydr Polym 67(1):97–103
Zhao X, Cheng K, Liu D (2009) Organosolv pretreatment of lignocellulosic biomass for enzymatic hydrolysis. Appl Microbiol Biotechnol 82(5):815–827
Zheng YZ, Lin HM, Tsao GT (1995) Supercritical carbon-dioxide explosion as a pretreatment for cellulose hydrolysis. Biotechnol Lett 17:845–850
Zheng YZ, Lin HM, Tsao GT (1998) Pretreatment for cellulose hydrolysis bycarbon dioxide explosion. Biotechnol Prog 14:890–896
Zhi S, Yu X, Wang X, Lu X (2012) Enzymatic hydrolysis of cellulose after pretreated by ionic liquids/; focus on one-pot process. Energy Procedia 2012(14):1741–47
Zhua JY, Wang GS, Pan XJ, Gleisner R (2009) Specific surface to evaluate the efficiencies of milling and pretreatment of wood for enzymatic saccharification. Chem Eng Sci 64:474–485
Zhu SD, Yu ZN, Wu YX (2005) Enhancing enzymatic hydrolysis of rice straw by microwave pretreatment. Chem Eng Commun 192:1559–1566
Zhu SD, Wu YX, Chen QM, Yu N, Wang CW, Jin SW (2006) Dissolution of cellulose with ionic liquids and its application: a mini-review. Green Chem 2006(8):325–7
Zhu Z, Sathitsuksanoh N, Vinzant T, Schell DJ, McMillan JD, Zhang Y-HP (2009) Comparative study of corn stover pretreated by dilute acid and cellulose solvent-based lignocellulose fractionation: enzymatic hydrolysis, supramolecular structure, and substrate accessibility. Biotechnol Bioeng 103:715–724
Zhu JY, Pan X, Ronald S, Zalesny J (2010a) Pretreatment of woody biomass for biofuel production: energy efficiency, technologies, and recalcitrance. Appl Microbiol Biotechnol 87:847–857

Further Reading

Bak JS, Ko JK, Han YH (2009) Improved enzymatic hydrolysis yield of rice straw using electron beam irradiation pretreatment. Bioresour Technol 100:1285–1290
Esteghalian A, Hashimoto AG, Fenske JJ, Penner MH (1997) Modelling and optimization of dilute-sulfuric-acid pretreatment of corn stove, poplar and switchgrass. Bioresour Technol 59:129–136
Li Z, Jiang Z, Fei B (2012c) Ethanol organosolv pretreatment of bamboo for efficient enzymatic saccharification. BioResources 7(3):3452–3462
Pye EK, Lora JH (1991) The AlcellTM process: a proven alternative to kraft pulping. Tappi J 74(3):113
Wang Q, Wu Y, Zhu S (2011b) Use of ionic liquids for improvement of cellulosic ethanol production. BioResources 6(1):1–2
Zhang Y-HP, Ding S-Y, Mielenz JR, Elander R, Laser M, Himmel M, McMillan JD, Lynd LR (2007c) Fractionating recalcitrant lignocelluloses at modest reaction conditions. Biotechnol Bioeng 97:214–223

Chapter 7
Ethanol Markets

Abstract Ethanol markets are discussed in this chapter. The ethanol market is influenced by national and state energy policies with some states mandating ethanol content, sometimes without mandatory labelling of the ethanol content. State and federal tax incentives have a strong influence on the market and production of ethanol in few countries.

Keywords Biofuels · Bioethanol · E10 · E20 · E50 · E85 · E100 · Fuel cell · E diesel

The ethanol market is driven by a number of factors including oil prices, national energy and foreign policy, federal and state tax incentives, and technological developments in the future (Bajpai 2007). As expected, as oil prices increase, there is often a shift in the demand for ethanol as a biofuel. However, the market is often limited by the amount that can be blended with gasoline for conventional engines and the limited availability and use of FlexFuel vehicles. The limited availability of distribution points for ethanol-based fuel is also a consideration. The development of biofuel vehicles must go hand in hand with the development of the fuel distribution system. The ethanol market is also influenced by national and state energy policies with some states mandating ethanol content, sometimes without mandatory labelling of the ethanol content. State and federal tax incentives have a strong influence on the market and production of ethanol in the United States. Needless to say, many corn-to-ethanol plants would not be profitable without these incentives. It is expected that technological changes, especially the use of different biomass feedstocks such as wood and lignocellulosic materials, will have a significant effect on the market. Even with these considerations, the world ethanol production is expected to increase significantly with the United States and Brazil being the leading producers in the world, perhaps reaching as high as 28 billion gallons. Emerging market and production in Cuba, Asia, and Latin America will also contribute to the growth of the ethanol market. It is expected that much of this growth in ethanol production will need to be in the form of cellulosic ethanol so that the agricultural and food market (with respect to corn) is not significantly disrupted.

P. Bajpai, *Developments in Bioethanol*, Green Energy and Technology,
https://doi.org/10.1007/978-981-15-8779-5_7

Table 7.1 Common ethanol fuel mixtures

E5—maximum 5% anhydrous ethanol; minimum 95% gasoline Western Europe
E10—maximum 10% anhydrous ethanol; minimum 90% gasoline USA today; Western Europe in near future
E15—maximum 15% anhydrous ethanol; minimum 85% gasoline USA EPA approval Cars >2000
E25—maximum 25% anhydrous ethanol; minimum 75% gasoline Brazil
E85—maximum 85% anhydrous ethanol; minimum 15% gasoline USA/ Europe
E100—100% Brazilian hydrous ethanol (contains on average 5.3 volume water) Brazil

Currently, the largest market for ethanol is transportation fuel for passenger vehicles. EPA under the "Gasohol Waiver" which became effective on December 16, 1978 allowed addition of ethanol to commercial gasoline. This allowed up to 10% by volume ethanol. The use of ethanol in transportation fuel has been gradually growing since the 1980s. The industry in United States in 1981 produced 200 million gallons yearly and continued stable growth to 15.8 billion gallons in 2017. As the production of ethanol increased, it was blended with gasoline to replace the octane enhancer's lead, benzene, toluene, and xylene as they were being removed from the gasoline supply because of toxicity issues. Ethanol has a blending octane of 113 and is mostly used in producing regular type of octane gasoline from sub-octane base stocks or increasing regular octane fuels to the mid-octane level. Adding ethanol to gasoline for boosting octane is an option to more harsh refining operations makes ethanol one of the most cost-effective octane enhancers available to the refiner and blender today (https://ethanolrfa.org/wp-content/uploads/2018/07/Fuel-Ethanol-Industry-Guidelines-Specifications-2018.pdf).

Several common ethanol fuel mixtures are being used all over the world (Table 7.1). "Pure hydrous or anhydrous ethanol can be used in internal combustion engines (ICEs) if the engines are designed or modified for that purpose, and used only in automobiles, light-duty trucks and motorcycles. Anhydrous ethanol can be blended with gasoline (petrol) for use in gasoline engines, but with high ethanol content only after minor engine modifications. Ethanol fuel mixtures have "E" numbers which describe the percentage of ethanol fuel in the mixture by volume, for example, E85 is 85% anhydrous ethanol and 15% gasoline. Low-ethanol blends, from E5 to E25, although internationally the most common use of the term refers to the E10 blend" (www.secret-bases.co.uk).

In passenger vehicles, ethanol can be used in any combination of blends. It can be used, up to 10% without making any modification in the vehicle. This includes E-10, oxygenated fuel, and reformulated gasoline (RFG). E-10 is commonly referred to as "gasohol". Higher blends, containing up to 85% ethanol such as E-85, can only

Table 7.2 Advantages and disadvantages of E85

Advantages	Disadvantages
• Domestically produced, reducing use of imported petroleum • Lower emissions of air pollutants • More resistant to engine knock • Added vehicle cost is very small	• Can only be used in flex-fuel vehicles • Lower energy content, resulting in fewer miles per gallon • Limited availability • Currently expensive to produce

Based on http://www.dingscompletecarcare.com/Go_Green.htm

be used in modified vehicles. All the three major automobile manufacturers are manufacturing E-85 flexible fueled vehicles. These are modified to run on ethanol blends up to 85%. Table 7.2 shows advantages and disadvantages of E85. There are also additional niche markets for ethanol-blended fuels (Launder 1999).

Ethanol, with its high-octane property, has established its role as an octane enhancer. Blending ethanol at 10% raises the gasoline's octane level by an average of three octane points. As an octane enhancer, ethanol had to compete with other octane enhancers. In the early 1980s, methanol blends gained a short popularity but soon disappeared from the market due to public health risks associated with methanol's toxicity as well as to its undesirable corrosive effects on engines and pipelines. While methanol blends did not experience a wide market success, a methanol-derived ether called MTBE, also used to raise octane content, became popular as a blending agent. MTBE, first produced in 1979 in the United States and Europe, was developed primarily as an octane enhancer by combining isobutylene and methanol. Unlike ethanol, MTBE can easily be blended with gasoline at the refinery and transported by pipeline

7.1 Transportation Fuel

Countries like United States and Brazil have long promoted domestic ethanol production. Ethanol/gasoline blends were promoted as an environmentally driven practice, in the United States initially as an octane enhancer to replace lead. Ethanol is also used as an oxygenate in clean-burning gasoline to reduce vehicle exhaust emissions. In United States ethanol supplies today account for about 1% of the highway motor vehicle fuel market, in the form of a gasoline blending component. At present, most of this ethanol is used in a 10% blend with gasoline which is commonly referred to as "gasohol," a term which is being replaced with "ethanol/gasoline blends" or "E10." In some areas, lower percentage blends, containing 5.7 or 7.7% ethanol are also being used to match to air quality regulations affecting the oxygen content of reformulated gasoline (CFDC 1999). The 5.7% blend is California's formulation used to meet a 2% by weight federal oxygenate requirement in Phase II gasoline.

5% bioethanol blend does not require any engine modification and is covered by vehicle warranties. Combined with gasoline, ethanol increases octane levels while also promoting more complete fuel burning that reduces harmful exhaust emissions such as carbon monoxide and hydrocarbons (IEA 2004, 2006). In addition to ethanol/gasoline blend markets, ethanol has other motor fuel applications including:

(1) Use as E85, 85% ethanol and 15% gasoline
(2) Use as E100, 100% ethanol with or without a fuel additive
(3) Use in Oxydiesel, typically a blend of 80% diesel fuel, 10% ethanol and 10% additives and blending agents.

There are also smaller niche markets such as fuel cell applications, E diesel (a cleaner burning diesel fuel containing up to 15% ethanol), aviation etc. where ethanol can be utilized (Launder 1999). Ethanol is used in the chemical industry for making variety of basic and intermediate chemicals. Bioethanol could be of great economic and environmental interest in developing countries especially for cooking and lighting as a substitution for LPG (particularly for remote locations). Power and heat segment may be much larger market than the transport for bio ethanol.

There are several common ethanol fuel mixtures in use around the world. The use of pure hydrous or anhydrous ethanol in ICE is only possible if the engine is designed or modified for that purpose. Anhydrous ethanol can be blended with petrol in various ratios for use in unmodified gasoline engines, and with minor modifications can also be used with a higher content of ethanol. Ethanol fuel mixtures have "E" numbers which describe the percentage of ethanol in the mixture by volume, for example, E85 is 85% anhydrous ethanol and 15% gasoline. Low ethanol blends, from E5 to E25, are also known as *gasohol*, though internationally the most common use of the term gasohol refers to the E10 blend. Blends of E10 or less are used in more than twenty countries around the world, led by the United States, where ethanol represented 10% of the U.S. gasoline fuel supply in 2011. Blends from E20 to E25 have been used in Brazil since the late 1970s (ANFAVEA 2006). E85 is commonly used in the United States and Europe for flexible fuel vehicles. Hydrous ethanol or E100 is used in Brazilian neat ethanol vehicles and flex-fuel light vehicles and in hydrous E15 called hE15 for modern petrol cars in Netherlands. Table 7.3 shows Ethanol blends used in Brazil.

The Energy Policy Act of 2005 established a Renewable Fuels Standard (RFS). This standard requires the use of 4.0 billion gallons of renewable fuels in 2006, increasing each year to 7.5 billion gallons in 2012. Most of this requirement is expected to be met with ethanol. In United States, approximately 3.4 billion gallons of ethanol were consumed in 2004.

Ethanol can be used as a replacement for MTBE in light of water contamination and health concerns. Unlike Ethanol, MTBE is highly soluble in water and travels easily and swiftly to ground and surface water supplies. Even small amount of methanol either swallowed or absorbed through the skin is very harmful. It can cause blindness, permanent neurological damage, and death. The potential health hazards from use of MTBE was also documented in a report from the University of California-Davis titled "Health & Environmental Assessment of MTBE" which

Table 7.3 Ethanol blends used in Brazil

Year	Ethanol blend
1987–88	E22
1993–98	E22
2000	E20
2001	E22
2003	E20–25
2004	E20
2005	E22
2006	E20
2007	E23–25
2008	E25
2009	E25
2010	E20–25
2011	E18–E25
2012	E20
2013	E20
2014	E25
2015	E 27
2019	E 27

Based on http://en.goldenmap.com/Common_ethanol_fuel_mixtures
https://www.transportpolicy.net/standard/brazil-fuels-biofuels/
https://apps.fas.usda.gov/newgainapi/api/report/downloadreportbyfilename?filename=Biofuels%20Annual_Sao%20Paulo%20ATO_Brazil_8-9-2019.pdf

concluded that there are significant risks and costs associated with water contamination due to the use of MTBE. Researchers also found MTBE in over 10,000 groundwater sites in California. The report supports the use of ethanol in place of MTBE stating that the use of ethanol as an oxygenate would result in much lower risk to water supplies, lower water treatment costs in the event of a spill, and lower monitoring costs.

7.1.1 E10 or Less

E10 is sometimes called *gasohol*. It is a fuel mixture of 10% anhydrous ethanol and 90% gasoline that can be used in the ICE of most modern automobiles and light duty vehicles and without need for any modification on the engine or fuel system. E10 blends are typically rated as 2–3 octane higher than regular gasoline and are approved for use in all new United States automobiles, and are mandated in some areas for emissions and other reasons. The E10 blend and lower ethanol content mixtures have been used in several countries, and its use has been primarily driven by the several world energy crises that have taken place since the 1973 oil crisis (EAIP 2001).

Other common blends include E5 and E7. These concentrations are generally safe for recent engines that run on pure gasoline. As of 2006, mandates for blending bioethanol into vehicle fuels had been enacted in at least 36 states/provinces and 17 countries at the national level, with most mandates requiring a blend of 10–15% ethanol with gasoline.

E10 and other blends of ethanol are considered to be useful in decreasing United States dependence on foreign oil, and can reduce CO emissions by 20–30% under the right conditions. Although E10 does decrease emissions of CO and greenhouse gases such as carbon dioxide by an estimated 2% over regular gasoline, it can cause increases in evaporative emissions and some pollutants depending on factors like the age of the vehicle and weather conditions. E10 is also commonly available in the Midwestern United States. E10 has also been mandated for use in all standard automobile fuel in the state of Florida by the end of 2010. Due to the phasing out of MTBE as a gasoline additive and mainly due to the mandates established in the Energy Policy Act of 2005 and the energy independence and security Act of 2007, ethanol blends have increased throughout the US, and by 2009, the ethanol market share in the United States gasoline supply reached almost 8% by volume. The Tesco chain of supermarkets in the UK have started selling an E5 brand of gasoline marketed as 99 RON super-unleaded. Its selling price is lower than the other two forms of high-octane unleaded on the market, Shell's V-Power (99 RON) and BP's Ultimate (97 RON). Many petrol stations throughout Australia now also sell E10, typically at a few cents cheaper per litre than regular unleaded.

7.1.2 E15

In October 2010 the EPA granted a waiver to allow up to 15% of ethanol blended with gasoline to be sold only for cars and light pickup trucks with a model year of 2007 or later, representing about 15% of vehicles on the United States roads. In January 2011 the waiver was expanded to authorize use of E15 to include model year 2001 through 2006 passenger vehicles. The EPA also decided not to grant any waiver for E15 use in any motorcycles, heavy-duty vehicles, or non-road engines because current testing data does not support such a waiver. According to the Renewable Fuels Association the E15 waivers now cover 62% of vehicles on the road in the United States, and the ethanol group estimates that if all 2001 and newer cars and pickups were to use E15, the theoretical blend wall for ethanol use would be approximately 17.5 billion gallons per year. EPA is still studying if older cars can withstand a 15% ethanol blend.

In October 2018, the EPA began the rulemaking process to allow production of E15 (conventional fuel blended with up to 15% of ethanol) all year long. Currently, production of E15 is banned in the summer months—from June 1 to September 15—for environmental concerns, particularly because it is linked to an increased presence of smog when blended with gasoline. The industry believes that allowing blending of E15 year-round, combined with increased promotion and marketing of the product in the country, will increase domestic demand and encourage retailers to invest in the infrastructure necessary to handle higher blends of ethanol. The United States Department of Agriculture is also incentivizing investment in infrastructure for E15 and higher ethanol blends through the Biofuel Infrastructure Partnership, which combines federal, state, and private sector funding (https://www.usitc.gov/publications/332/executive_briefings/ebot_angelica_marrero_where_is_u.s._ethanol_going_pdf.pdf).

7.1.3 E20, E25

These blends have been widely used in Brazil since the late seventies. E20 contains 20% ethanol and 80% gasoline, while E25 contains 25% ethanol. As a response to the 1973 oil crisis, the Brazilian government made mandatory the blend of ethanol fuel with gasoline, fluctuating between 10% to 22% from 1976 until 1992. Due to this mandatory minimum gasoline blend, pure gasoline is no longer sold in Brazil. A federal law was passed in October 1993 establishing a mandatory blend of 22% anhydrous ethanol (E22) in the entire country. This law also authorized the Executive to set different percentages of ethanol within pre-established boundaries, and since 2003 these limits were fixed at a maximum of 25% (E25) and a minimum of 20% (E20) by volume. Since then, the government has set the percentage on the ethanol blend according to the results of the sugarcane harvest and ethanol production from sugarcane, resulting in blend variations even within the same year.

The mandatory blend was set at 25% of anhydrous ethanol (E25) by Executive Decree since July 1, 2007, and this has been the standard gasoline blend sold throughout Brazil most of the time as of 2011. All Brazilian automakers have adapted their gasoline engines to run smoothly with this range of mixtures. Some vehicles might work properly with lower concentrations of ethanol, however, with a few exceptions, they are unable to run smoothly with pure gasoline which causes engine knocking, as vehicles traveling to neighboring South America countries have demonstrated. Flexible fuel vehicles, which can run on any mixed of gasoline E20-E25 up to 100% hydrous ethanol (E100 or hydrated ethanol) ratios, were first available in mid 2003. In July 2008, 86% of all new light vehicles sold in Brazil were flexible-fuel, and only two carmakers build models with a flex-fuel engine optimized to operate with pure gasoline. A state law approved in Minnesota in 2005 mandates that ethanol comprise 20% of all gasoline sold in this American state beginning in 2013. Successful tests have been conducted to determine the performance under E20 by current vehicles and fuel dispensing equipment designed for E10.

7.1.4 E70, E75

When the vapor pressure drops below 45 kPa in the ethanol blend, fuel ignition cannot be assured on cold winter days. This limits the maximum ethanol blend percentage during the winter months to E75. E70 and E75 are the winter blends used in the United States and Sweden for E85 flexible fuel vehicles during the cold weather, but still sold at the pump labeled as E85. The seasonal reduction of the ethanol content to a E85 winter blend is mandated to avoid cold starting problems at low temperatures. In the United States, this seasonal reduction of the ethanol content to E70 applies only in cold regions, where temperatures fall below 32 °F during the winter. In Sweden, all E85 flexible fuel vehicles use a E75 winter blend. This blend was introduced since the winter 2006–07 and E75 is used from November until March. All E85 flex vehicles require an engine block heater to avoid cold starting problems. The use of this device is also recommended for gasoline vehicles when temperatures drop below −23°. Another option when extreme cold weather is expected is to add more pure gasoline in the tank, thus reducing the ethanol content below the E70 winter blend, or simply not to use E85 during extreme low temperature.

7.1.5 E85

E85 is generally the highest ethanol fuel mixture found in United States and Europe, as this blend is the standard fuel for Flexible-fuel vehicles. "Currently, the use of E85, a mixture of 85% ethanol and 15% gasoline, for flexible fuel vehicles (FFV) or variable fuel vehicles (VFV), has become common; and there are various manufacturers (such as Ford, General Motors, Chrysler Corporation, etc.) offer vehicles (estimated at over 7 million FFVs on the roads by 2009) that are capable of operating on 100% gasoline, or E85, or any mixture of the two. Ethanol gasoline blends of various amounts are commonly used in different countries worldwide; e.g. Australia (officially 10%), Brazil (up to 25%), Canada (10%), Sweden (5%) and the USA (up to 10%). However, in the United States, automotive manufacturers agreed that the use of gasoline with up to 10% ethanol will not affect the warranties of their vehicles (Egebäck et al. 2005; Kheiralla et al. 2017; Launder 2001; soeagra.com)".

The octane rating of this mixture is about 105, which is very much lower in comparison to pure ethanol but still higher than normal gasoline (87–95 octane, depending on country). The 85% limit in the ethanol content was set for reducing ethanol emissions at low temperatures and avoiding cold starting problems in the cold weather, at temperatures lesser than 11 °C. A further reduction in ethanol is used during the winter in regions where temperature falls below 0 °C and this blend is called Winter E85, as the fuel is being sold using the E85 label. A winter blend of E70 is mandated in some regions in the United States, whereas Sweden mandates E75. In October 2010 there were nearly 3000 E85 fuel pumps in Europe, led by Sweden with 1699 filling stations. The United States had 2414 public E85 fuel pumps located in 1701 cities by October 2010, mostly found in the Midwest (www.flowersplantsgardening.com).

Ethanol produced by cellulosic raw materials offers significant carbon dioxide emissions reductions compared with fossil-based transport fuels for internal combustion engine (ICE) passenger vehicles, and also for trucks and buses when used as ED95 (95% fuel ethanol with lubricants and additives). Although regular vehicles can accommodate ethanol at low blend rates, carbon dioxide emissions reductions are maximised when it is used at high blend shares or unblended in flexible-fuel vehicles. Higher cellulosic ethanol production would also provide the additional benefit of restraining agricultural residue-burning in fields, which worsens the air quality (https://www.iea.org/reports/tracking-transport-2019/transport-biofuels).

7.1.6 E100

E100 is pure ethanol fuel. Hydrous ethanol as an automotive fuel has been widely used in Brazil since the late seventies for neat ethanol vehicles and more recently for flexible fuel vehicles. The ethanol fuel used in Brazil is distilled close to the azeotrope mixture of 95.63% ethanol and 4.37% water which is approximately 3.5% water by volume. The azeotrope is the highest concentration of ethanol that can be achieved via distillation. The maximum water concentration according to the ANP specification is 4.9 vol%. The E nomenclature is not adopted in Brazil, but hydrated ethanol can be tagged as E100 meaning that it does not have any gasoline, because the water content is not an additive but rather a residue from the distillation process. However, straight hydrous ethanol is also called E95 by some authors. The first commercial vehicle capable of running on pure ethanol was the Ford Model T, produced from 1908 through 1927. It was fitted with a carburetor with adjustable jetting, allowing use of gasoline or ethanol, or a combination of both. At that time, other car manufactures also provided engines for ethanol fuel use. Thereafter, and as a response to the energy crisis the first modern vehicle capable of running with pure hydrous ethanol (E100) was launched in the Brazilian market, the Fiat 147 after testing with several prototypes developed by the Brazilian subsidiaries of Fiat, Volkswagen, General Motors and Ford. Since 2003, Brazilian newer flexible fuel vehicles are capable of running on pure hydrous ethanol ethanol (E100) or blended with any combination of E20 to E25 gasoline (a mixture made with anhydrous ethanol), the national mandatory blend.

E100 imposes a limitation on normal vehicle operation as ethanol's lower evaporative pressure (as compared to gasoline) causes problems when cold starting the engine at temperatures below 15 °C. For this reason both pure ethanol and E100 flexible fuel vehicles are built with an additional small gasoline reservoir inside the engine compartment to help in starting the engine when cold by initially injecting gasoline. Once started, the engine is then switched back to ethanol. An improved flex engine generation was developed to eliminate the need for the secondary gas tank by warming the ethanol fuel during starting, and allowing flex vehicles to do a normal cold start at temperatures as low as −5 °C the lowest temperature expected anywhere in the Brazilian territory. The Polo E-Flex, launched in March 2009, was the first flex fuel model without an auxiliary tank for cold start. The higher fuel efficiency of E100 in high performance race cars resulted in Indianapolis 500 races in 2007 and 2008 being run on 100% fuel grade ethanol.

7.2 Niche Markets

There are also many smaller niche markets where ethanol can be utilized. They are considered niche markets because they are much smaller than the general transportation fuel market. However, they can still create a considerable demand, especially for a much smaller fuel-producing industry like ethanol. Opportunities exist as ethanol-blended diesel fuel, as a hydrogen source for fuel cells, as an aviation fuel, snowmobiles, and other off-road vehicles, boats, and other personal watercraft, and small engine equipment (RFA 2004, 2005, 2006a, b, c).

7.2.1 Fuel Cells

One of the newest markets being looked at for ethanol use is fuel cells. Fuel cells create electricity by combining hydrogen and oxygen. Fuel cells are more energy efficient than the internal combustion engine. Although fuel cells are considered a new technology for vehicles and other applications, they have actually been around for quite awhile. In fact, the basic configuration and idea for fuel cells was discovered in 1839. However, the commercial possibilities of fuel cells weren't realized till the 1960s when NASA began using fuel cells in spacecraft to provide power. In 1998, Chicago became the first city to use pure hydrogen-powered fuel-cell buses. Currently, all big three automobile manufacturers, as well as many foreign auto manufacturers, are developing fuel-cell vehicle prototypes. Fuel cells could operate on hydrogen from a number of sources including ethanol, methanol, gasoline, and perhaps, at some point, even water. Ethanol is expected to play an important role as a fuel cell fuel supplying the hydrogen for fuel cell operation (RFA, 2005a). Ethanol is a clean, renewable, hydrogen source that provides huge benefits in reducing the greenhouse gases (GHG) that contribute to climate change (Fuel Cells 2000). Nuvera Fuel Cells, formerly Epyx, has developed a fuel cell processor capable of converting ethanol, methanol, and gasoline to hydrogen. They have found that Ethanol provides higher efficiencies, fewer emissions and better performance than other fuel sources, including gasoline. Ethanol is also much less corrosive then methanol. Ethanol use in fuel cells is a great opportunity for consumers, the environment and the ethanol industry (RFA 2000a, b, c). The RFA has a Fuel Cell Task Force that continues to follow fuel cell developments and to position ethanol for a role in fuel cell. In United States, the "Hydrogen Fuel Initiative" supports research and commercialization of fuel cells for automobiles and power generation. The first commercial demonstration of ethanol's potential to produce hydrogen to power a fuel cell is underway at Aventine Renewable Energy, Inc.'s Pekin, Illinois ethanol plant. The 13 kW stationary fuel cell system is generating power for the plant's visitor center and additional energy for the plant. The project is a partnership with the United States Department of Energy, Caterpillar, Inc., Nuvera Fuel Cells, the State of Illinois, Renewable Fuels Association and Illinois Corn Growers Association.

Fuel cells work very similarly to batteries except they can run continuously as long as fuel is supplied (Fuel Cells 2000). Benefits of fuel cells use in the transportation sector include increased energy efficiency, a tremendous decrease in emissions, less vehicle maintenance and the ability to achieve up to 80 mpg. Fuel cells create electricity by combining hydrogen and oxygen. They could eventually be used to supply power to homes, vehicles, and small electronic devices.

Two major hurdles in the development of fuel cells for the transportation sector are storing hydrogen onboard a vehicle and developing a hydrogen refuelling infrastructure. One way around these obstacles is to create a process to extract hydrogen from already established fuels such as gasoline and ethanol. Benefits of using ethanol versus other fuels include lower emissions and higher efficiency. However, extracting hydrogen from ethanol requires higher temperatures than when using other fuels, such as methanol, and some argue that this may hinder ethanol use in fuel cells. Ethanol may make up for this shortcoming by increasing the amount of energy over methanol (25% higher), which allows for a longer vehicle range. Ethanol is also much less corrosive then methanol

7.2.2 E Diesel

Another market for ethanol considered to be in the developmental stage is the use of ethanol blends in diesel fuel (Launder 1999). These blends range from 7.7 to 15% ethanol, and 1 to 5% special additives that prevent the ethanol and diesel from separating at very low temperatures or if water contamination occurs. Ethanol blended diesel could provide a considerable increase in demand for ethanol as diesel vehicles in the United States consume approximately 50 billion gallons a year. Demonstrations are currently being conducted on the use of ethanol-blended diesel in heavy-duty trucks, buses, and farm machinery.

The use of ethanol-blended diesel has similar disadvantages to other ethanol-blended fuel, including decrease in gas mileage and increase in cost compared to regular diesel. Advantages include a decrease in emissions and in demand for imported petroleum. Presently E diesel fuels are considered experimental and is being developed by many companies. Blends containing up to 15v% ethanol, blended with standard diesel and a proprietary additive, are called E diesel fuels. A number of fleet demonstrations have been completed with favorable results and some controlled testing has also been completed. Results show that E diesel blends reduce certain exhaust emissions, especially particulates, in certain diesel applications and duty cycles. However, additional testing is needed to assess compatibility of E diesel blends with various fuel system parts and also to determine the long range effects on engine durability. Additional emissions tests are also needed to more accurately quantify the emissions profiles of various E diesel blend levels. The RFA is working closely with industry to address the research needs (RFA 2005; Vaughn 2000). Most E diesel blends have properties similar to standard diesel fuel, or can be modified to be similar through the use of additives. One of the important difference is the lower flash point of E diesel blends. E diesel blends are designated as Class I whereas diesel fuel is designated as a Class II flammable liquid.

7.2.3 Aviation

A more near-term niche market for ethanol is aviation. Aviation fuel only accounts for about 5% of the total transportation fuel use in the United States it still represents a sizeable market for ethanol consumption (approximately 400 million gallons per year). Currently, ethanol is used in some small engine crafts and by experimental aircraft pilots (Caddet 1997; Higdon 1997; Bender 2000). Research on the use of ethanol in aviation started more than 20 years ago. In March 2000, FAA certified AGE-85 (Aviation Grade Ethanol) which is a high-performance fuel that may be used in any piston engine aircraft. It contains approximately 85% ethanol, along with light hydrocarbons and biodiesel fuel. AGE-85 is specifically blended for cold starting and good mixture balance. AGE is unleaded, burns cleaner, has lower exhaust emissions and is more environmentally friendly than traditional aviation fuels. The ethanol in AGE-85 prevents carburetor and fuel line icing, and provides excellent detonation margins. Some constraints in using AGE-85 include its current lack of availability and decreased range (5–15%) due to its lower BTU content. Probably the most formidable constraint to ethanol being excepted in aviation is some opposition within the field itself. The Aircraft Owners and Pilots Association (AOPA) and Cessna have stated that any new unleaded fuel being considered for aircraft should be able to be used without aircraft modifications. They also are doubtful about the ability and willingness of airports to put up a new refueling infrastructure to accommodate a fuel such as AGE-85. Indeed, pilots would most likely have to secure funding or put up the money themselves to install E-85 refueling equipment-at least until widespread use significantly increases demand (Launder 1999).

There is also the potential for ethanol to be used in a biodiesel blend and/or ETBE in turbine aircraft, which would not require aircraft modifications. A biodiesel blend would be 20% biodiesel-80% regular diesel and could reduce particulate emissions by 80%. ETBE contains about 43% ethanol and isobutylene a gas/petroleum derivative. This could also prove to be a considerable market for ethanol as turbine aircraft used approximately 16.4 billion gallons of fuel in 1997.

Presently, ethanol is used in some small engine crafts and by experimental aircraft pilots. The largest general aviation market in the world is the United States and the second largest is Brazil. Pressure is being put by EPA on the aviation sector to take the lead out of their fuel. It appears likely that Avgas (the leaded aviation standard fuel) will need to be replaced with an unleaded fuel in the near future and that ethanol could be the replacement fuel.

7.2.4 Snowmobiles

Another potential market for the use of ethanol is snowmobiles. Snowmobiles could provide a considerable market for ethanol in United States. There are a total of over 2.3 million snowmobiles registered in the United States. The snowmobile manufacturers approve the use of ethanol blends of up to 10%. The use of ethanol in snowmobiles greatly reduces emissions. In the United States, several million snowmobiles are registered. This is a significant factor to consider as a snowmobile with a conventional 2-stroke engine emits 36 times more carbon monoxide and 98 times more hydrocarbons than an automobile. In fact, snowmobiles can create serious environmental and health risks that many national parks are considering limiting or banning the use of snowmobiles within the parks. Using a blend of 10% ethanol also eliminates the need for the use of gas-line antifreeze and removes and prevents deposit build up in the fuel tank (Launder 1999).

7.2.5 Boats/Marine

Ethanol has been shown to be a viable alternative to gasoline for use in recreational boat engines, due to better environmental performance as a fuel over gasoline (Launder 1999). Many marine/boat manufacturers approve the use of ethanol blends of up to 10%. Study conducted by Dambach et al. (2004) has proven the ability to retrofit an engine with minimal modifications and lose little in the way of performance. The use of alcohol based fuels in older boats may cause some problems as the older boats may not have alcohol compatible parts and therefore certain parts could be degraded with the use of ethanol blended fuels. Ethanol blends of up to 10% can also be used in boats and personal watercraft. This could also prove to be a considerable size market United States. Many marine/boat manufacturers approve the use of ethanol-blended fuels. However, some boat manufacturers do have cautionary language on the use of alcohol based fuels in older boats as they may not have alcohol compatible parts and therefore certain parts could be degraded with the use of ethanol blended fuels.

7.2.6 *Small Engine Equipment*

Ethanol blends can also be used in small engines such as lawnmowers, chainsaws, and weed trimmers (Launder 1999). The benefits of using ethanol would again be decreased emissions. The Portable Power Equipment Manufacturers Association has conducted reformulated fuel research for chainsaws, weed trimmers and other hand held equipment and observed no operating problems with equipment when using reformulated gasoline. Study conducted by California Environmental Protection Agency has shown that a gasoline-powered lawnmower run for an hour puts out about the same amount of smog-forming emissions as 40 new automobiles run for an hour. They also found that a chain saw operated for 2 h is able to generate the same amount of emissions as ten automobiles driven 250 miles.

References

ANFAVEA—National Association of Automobile Fabricants (2006) Brazilian automotive industry yearbook, p 167 (in Portuguese)
Bajpai P (2007) Bioethanol. PIRA Technology Report, Smithers PIRA, UK
Bender B (2000) Ethanol: aviation fuel of the future. Michigan Ethanol Workshop Presentation
Caddet (1997) Ethanol as an aviation fuel. Caddet renewable energy. Technical Brochure No. 51. http://lib.kier.re.kr/caddet/retb/no51.pdf
Clean Fuels Development Coalition (CFDC) (1999) Fuel ethanol fact book. Bethesda
Dambach E, Han A, Henthorn B (2004) Ethanol as fuel for recreational boats. ENGS 190/ENGG 290 final report. http://www.dartmouth.edu/~ethanolboat/Ethanol_Outboard_Final_Report.pdf
EAIP (Earth Policy Institute) (2001) Setting the ethanol limit in petrol. Environment Australia issues paper
Egebäck K, Henke M, Rehnlund B, Wallin M, Westerholm R (2005) Blending of ethanol in gasoline for spark ignition engines—problem inventory and evaporative measurements. MTC AB Report No. MTC 5407, ISSN: 1103-0240, ISRN: ASB-MTC-R—05/2—SE, p 133
Fuel Cells (2000) Benefits of fuel cells. http://216.51.18.233/fcbenefi.html
Higdon D (1997) Treaty calls for Great Lakes ban on leaded fuel. General Aviation News & Flyer
IEA (International Energy Agency) (2004) Biofuels for transport—an International perspective. Paris. http://www.iea.org/textbase/nppdf/free/2004/biofuels2004.pdf
IEA (International Energy Agency) (2006) Medium term oil market report. Paris. http://www.iea.org
Kheiralla AF, Tola E, Bakhit JM (2017). Performance of ethanol-gasoline blends of up to E35 as alternative Automotive fuels. Adv Biores 8(5):130–140
Launder K (1999) Opportunities and constraints for ethanol-based transportation fuels. Lansing: State of Michigan, Department of Consumer and Industry Services, Biomass Energy Program. http://www.michigan.gov/cis/0,1607,7-154-25676_25753_30083-141676–,00.html
Launder K (2001) From promise to purpose: opportunities and constraints for ethanol based transportation fuels. M.Sc. Thesis, Department of Resource Development, Michigan State University, USA, p 53
RFA—Renewable Fuels Association (2000a) Ethanol as a renewable fuel source for fuel cells. http://www.ethanolRFA.org/fuelcells.htm
RFA—Renewable Fuels Association (RFA) (2000b) President issues executive order to green the federal fleet on Earth Day. Ethanol Report, 27 Apr 2000

RFA—Renewable Fuels Association (RFA) (2000c) Ethanol—its use in gasoline: expected impacts and comments of expert reviewers. http://www.ethanolrfa.org/objects/documents/87/ethanol_tox_20001.pdf
RFA—Renewable Fuels Association (2004) Ethanol industry outlook: synergy in energy. http://www.ethanolrfa.org/objects/pdf/outlook/outlook_2004.pdf
RFA—Renewable Fuels Association (2005) Ethanol industry outlook: homegrown for the homeland. http://www.ethanolrfa.org/objects/pdf/outlook/outlook_2005.pdf
RFA—Renewable Fuels Association (2006a) Ethanol industry outlook: from niche to nation. www.ethanolrfa.org/objects/pdf/outlook/outlook_2006.pdf
RFA—Renewable Fuels Association (2006b) Fuel ethanol industry plants and production capacity. U.S
RFA—Renewable Fuels Association (2006c) Statistics data. http://www.ethanolrfa.org/industry/statistics/#C
Vaughn E (2000) RFA. International fuel ethanol workshop presentation. www.sentex.net/~crfa/crfanew.html

Chapter 8
Characteristics of Fuel Ethanol

Abstract Characteristics of fuel ethanol are discussed in this chapter. Ethanol is more reactive than hydrocarbon fuels, such as gasoline. Its molecular structure shows a polar fraction due to the hydroxyl radical and a non polar fraction in its carbon chain. Ethanol can be dissolved in both gasoline (non polar) and in water (polar). Due to its short carbon chain, the properties of ethanol polar fraction overcome the non polar properties). The formation of hydrogen bridges in ethanol molecule results in higher boiling temperature in comparison to that of gasoline.

Keywords Biofuels · Bioethanol · Fuel ethanol · High-octane fuel · Environmentally friendly · Ethanol based engine

Ethanol is more reactive than hydrocarbon fuels, like gasoline. The molecular structure of ethanol shows a polar fraction because of the hydroxyl radical and a non polar fraction in its carbon chain. Therefore ethanol can be dissolved in both gasoline and in water. Gasoline is non polar and water is polar. Because of its shorter carbon chain, the properties of polar fraction of ethanol overcome the non polar properties (Bajpai 2007). Hydrogen bridges are formed in ethanol molecule results in higher boiling temperature as compared to that of gasoline (Table 8.1).

Mixing of ethanol with gasoline has numerous effects. The heat output of the unleaded gasoline is increased by mixing ethanol. It produces more complete combustion which results in little lower emissions from hydrocarbons which are not burnt. When the concentrations of ethanol is higher, the fuel has more polar solvent type characteristics and has corresponding effects on carrying out fire suppression operations. But, at high concentration of ethanol, smallest amount of water will draw the ethanol from the blend away from the gasoline. Ethanol and gasoline have similar specific gravity. These two different fuels mix readily with minimum agitation, but the mixture is more of a suspension and not a true solution. Ethanol shows a greater affinity for water than it does for gasoline. In due course, without agitation, gasoline will float on a layer of an ethanol–water solution. The resulting ethanol–water solution is still flammable as the concentration of ethanol is still reasonably rich. Separation of phase may take place in fuel storage system where water is present. Mixing

P. Bajpai, *Developments in Bioethanol*, Green Energy and Technology, https://doi.org/10.1007/978-981-15-8779-5_8

Table 8.1 Properties of hydrous ethanol blend and gasoline-ethanol blend

Parameter	Hydrous ethanol	78% Gasoline 22% ethanol
Chemical structure	C2 H6.16 O1.08	C6.39 H13.60 O0.61
Carbon mass (%)	50.59	76.7
Hydrogen mass (%)	12.98	13.6
Oxygen mass (%)	36.42	9.7
Sulphur mass (%)	0	0.09
Self-ignition temperature (°C)	420	400
Temperature of vaporization (°C)	78	40–220
Heat of vaporization (kcal/kg)	105	237
Research octane number (RON)	–	106
Motor octane number (MON)	87	80
Vapour pressure (bar)	29	27.5
Laminar flame speed (m/s)	0.42	0.30
Density (kg/l) 0.74 0.81	0.81	0.74
Lower heating value (kcal/kg)	5970	9400
Stoichiometric air/fuel ratio	8.7	13.1

Based on Costa and Sodre (2010)

these fuels together changes the physical and chemical characteristics of the original fuels. But, the resulting changes may be not noticeable to emergency responders. One of the main differences in the blended fuel versus unblended gasoline is the visual difference in the characteristic of smoke and flame. When the content of ethanol is higher, the black smoke content and orange flame production is not much observable. These qualities may be masked by other substrates that may also be burning such as vehicle tires. Another noticeable difference of ethanol blended fuels under fire conditions is that when foam or water has been flowed on the burning product, the gasoline tends to burn off first, finally leaving the less volatile ethanol-water solution which may have no noticeable flame or smoke.

Table 8.2 shows key properties of petrol, ethanol (E-100), and 10% ethanol to petrol blend (E-10).

Table 8.2 Key properties of petrol, ethanol (E-100), and 10% ethanol to petrol blend (E-10) [5]

Properties	Petrol	E-100	E-10
Viscosity mm2/s	0.48–0.84	1.57 (20 °C)	0.53 (30 °C)
Flash Point °C	– 65	13	−40
RON °C	86.4–100	108.6–114	87.4–94
MON °C	80–98.8	89.7–112	86–99.9
Octane number	86–94	98–100	–
Cloud point °C	– 22	–	8
Pour point	(−17)–(−19)	–	−0
MJ/Kg	41.9–44.4	26–30	33.19–44.22

Based on Al-Mashhadani (2017), Fivgal et al. (2019)

Ethanol is less toxic than methanol – another alcohol used as fuel. The simple structure of ethanol molecule makes it suitable for spark ignition internal combustion engines operation. The high octane number of ethanol allows for higher compression ratios in comparison to gasoline fuelled engines (Table 8.1) (Costa and Sodre 2010).

One of the main advantages ethanol offers when compared to gasoline is its anti-knock performance, allowing its use in higher compression ratio engines. At high temperature, ethanol produces superior thermal efficiency due its higher heat of vaporization. As ethanol can burn richer fuel/air mixtures, it allows for higher engine power output in comparison to gasoline. However, due to its lower heating value, the use of ethanol instead of gasoline results in higher fuel consumption. Engine cold start is also a problem for ethanol, due to its low vapour pressure (Owen and Coley 1995).

Rasskazchikova et al. (2004) discussed the use of ethanol as high octane additive to automotive gasoline. The authors concluded that, despite the high cost, ethanol is the most promising octane-raising additive available. That is justified by the low toxicity, reduced environmental pressure when burning ethanol-containing fuel, and production from renewable raw material. Silva et al. (2005) evaluated the effect of ethanol and other additives on the anti-knock properties and Reid vapour pressure (RVP) of gasoline. Addition of ethanol up to 25% by volume in gasoline led to increased RVP and octane ratings.

Ethanol is used as an automotive fuel and can be used alone in specially designed engines, or blended with gasoline and used without any engine modifications. Motorboats, motorcycles, lawnmowers, chain saws etc. can all utilize the cleaner gasoline/ethanol fuel.

Ethanol is a high-octane fuel with high oxygen content. So, it allows the engine to more completely combust the fuel, resulting in fewer emissions. Ethanol has many advantages as an automotive fuel. Ethanol blends also reduce carbon monoxide emissions. When used in a correctly formulated fuel, ethanol can also reduce vehicle emissions which contribute to the formation of smog. It can also withstand cooler temperatures. Because of its low freezing point it can be used as the fluid in thermometers for temperatures below −40 °C, and for other low temperature purposes, such as for antifreeze in automobile radiators.

The key to better quality fuel is higher octane number as it provides higher torque, power and effectiveness. The **octane** rating of ethanol is 114. It is the highest quality fuel available in the market and keeps high compression engines of today to run efficiently.

Thanks to ethanol, there is no requirement to mix octane increasing additives which are deleterious to the environment like methyl tertiary-butyl ether (MTBE). The higher octane content of ethanol is also very important in making certain high compression engines to run smoothly and with record higher fuel prices during the last decade, more and more car manufacturers now a days are replacing large naturally aspirated engines with smaller displacement high compression turbocharged or supercharged engines. These engines offer better fuel economy. In view of this, ethanol will play an important role in developing these engines whereas fuels with higher proportion of ethanol will unlock the true potential of these engines.

Ethanol and detergents help to keep gummy deposits out of the fuel system of the car. When unleaded gasoline consisting of ethanol and detergents replaced leaded gasoline starting in the middle of the 1980s, some initial problems were faced as deposits were washed out from fuel systems and plugged filters. Now a days, the use of unleaded gasoline mixed with ethanol has reduced deposits in fuel systems, and all gasolines sold in the United States include detergents for keeping the fuel system clean (ethanolrfa.org)

Harmful emissions of tailpipe such as carbon monoxide, and other hydrocarbons and fine particulates found in car engine exhaust are reduced by using the renewable fuels.

Ethanol has a substantially higher heat of vaporation, provides cooling effect which helps in reducing engine knock.

Tests and fleet studies, performed by Amoco and the CRC showed no difference in vehicle performance in comparison to gasoline not blended with ethanol.

"Many automakers have produced flex-fuel vehicles (FFVs) since that can run on gasoline without ethanol, E85 (85% ethanol, 15% gasoline), or any combination of the two. There are now more than 20 million FFVs on the road and they are produced by GM, Ford, Chrysler, Jeep, Mercedes and others. These vehicles use modern technology to automatically adjust the vehicle to whatever blend is purchased. You can identify these vehicles by checking your owner's manual, checking the rear of your vehicle for a flex-fuel badge, visiting with your dealer, or checking the fuel filter door or gas cap for more information. FFVs must be identified in one of these ways, but each automaker is different" (https://ethanolrfa.org/consumers/ethanol-engines/#1553841684669-b1c8e2e7-fbb1).

The Energy Policy Act of 2005 established RFS, which mandates the use of ethanol and other renewable fuels in gasoline. Most of the fuel ethanol consumed in United States is E 10. A blend of 85% ethanol and 15% unleaded gasoline (E-85) is also being used. Currently, in United States approximately 50 million gallons of ethanol are made into E85. E85 is available at 1,000 locations in the United

States (including both public and private). The National Ethanol Vehicle Coalition estimates approximately 6 million FFVs on America's roads today as compared to approximately 230 million gasoline- and diesel-fueled vehicles. Most E85-capable vehicles are "flexible fuel vehicles" or FFVs (RFA 1999, 2001, 2006a, b, c, 2007a, b). This mixture has an octane rating of about 105. This is significantly lower from pure ethanol but still much higher than normal gasoline. The addition of a small amount of gasoline helps a conventional engine start when using this fuel under cold conditions. E85 does not always contain exactly 85% ethanol. In winter, particularly in colder climates, additional gasoline is added to facilitate cold start. E85 has been similar in cost to gasoline, but with the large oil price rises of 2005, it is being sold for as much as $0.70 less per gallon than gasoline which makes it highly attractive to the small but growing number of motorists with cars capable of burning it. Gasoline contains more energy, gallon for gallon than ethanol. One gallon of gasoline contains approximately 114,132 btu and ethanol contains 76,000 btu. Therefore E-85 contains approximately 27% less btu then 100% gasoline. However, when factoring in the fuel efficiency of gasoline and ethanol (which is more fuel efficient), it was found that E-85 mileage reduction is only about 10 to 15% instead of the 27% based strictly on energy content.

E-85 is environmentally friendly (Niven 2005). It has the highest oxygen content of any fuel available today, making it burn cleaner than ordinary gasoline. The use of E-85 reduces pollutants such as ozone and carbon monoxide and air toxins like benzene. E-85 cars perform well with significant reductions in emissions when compared to vehicles using ordinary unleaded gasoline. Reductions in two particularly troublesome pollutants—carbon monoxide and hydrocarbons are reduced significantly. Ethanol is one of only two liquid fuels available that combats global warming because of its raw material source. As corn grows, it converts carbon dioxide into oxygen. Auto makers are offering more flexible fuel vehicles. Purchase price of these vehicles has been comparable to the base price of gasoline models. Since E-85 is a cleaner burning fuel, it is expected that the life of a flexible fuel vehicle will be somewhat longer than that of a comparable gasoline vehicle.

The key component in a variable fuel vehicle is a sensor that determines the percentage of ethanol in the fuel. With the help of a computer, the vehicle automatically adjusts for best performance and emissions. Chrysler began offering E-85 minivans in the 1998 model year and Ford offered the Taurus and added Windstar and Ranger to the E-85 flexible fuel vehicles in the 1999 model year. Ford, GMC, Chevrolet and Daimler-Chrysler are now offering E-85 variable fuel vehicles.

The mileage decrease which occurs when operating a vehicle on E-85 has been debated. Ford reports an average of 16 MPG for E85 Taurus (based on city and highway driving) and 22 MPG for gasoline (Ford Motor Company 1998). Daimler-Chrysler reports a 27% range reduction when using E85 in their minivans (Chrysler Corporation 1997).

The color of ethanol fuel blends depends on the color of the gasoline in the blend. Blends may also have a gasoline-like odor. A gallon of E85 contains 27% less energy. The energy content of a gallon of ethanol is equal to 76,000 BTU as compared to about 115,000 BTU for a gallon of conventional gasoline. It follows that to replace the energy equivalent of a gallon of gasoline, we would need approximately 1.5 gallons of ethanol.

The average vehicle on the road today emits more than 600 lb of pollution into the air each year in United States. These harmful emissions include carbon monoxide, volatile organic compounds, particulate matter, oxides of nitrogen, and carbon dioxide. These emissions have significant health implications because they contribute to the amount of smog and carbon monoxide in air. Carbon monoxide emissions have also been implicated in global warming. One of the benefits of using E85 vehicles is a reduction in the amount of pollutants emitted into the air we breathe. The emissions control systems found on ethanol-powered vehicles manufactured today have been engineered to meet or exceed all federal and state emissions control regulations. Two types of emissions are released by E85 vehicles—exhaust and evaporative. Although compliance with federal and state regulations has already resulted in a decrease in exhaust emissions from gasoline-powered vehicles, ethanol-fueled vehicles can further reduce pollution from emissions by a modest amount. Most ethanol-fueled vehicles produce lower carbon monoxide and carbon dioxide emissions and the same or lower levels of hydrocarbon and non methane hydrocarbon emissions compared with gasoline-fueled vehicles. Nitrogen oxide emissions are about the same for ethanol and gasoline vehicles.Emissions resulting from fuel evaporation are a potential problem for any vehicle, regardless of the fuel. E85 has fewer highly volatile components than gasoline and so has fewer emissions resulting from evaporation.

Brazil has used ethanol blends since 1939. High oil prices in the 1970s prompted a government mandate in Brazil to produce vehicles fueled by pure ethanol in order to reduce dependence on foreign oil and provide value-added markets for its sugar cane producers. Today, there are several million ethanol-powered vehicles in Brazil that consume more than 4 billion gallons of ethanol annually.

Requirements in the Clean Air Act to make cleaner burning reformulated gasoline with lower volatility and fewer toxic components have increased interest in ethanol-based ethers such as ethyl tertiary butyl ether. It is a chemical compound produced by reacting ethanol and isobutylene. ETBE has superior physical and combustion characteristics to other ethers. They include: low volatility, high octane value, lower carbon monoxide and hydrocarbon emissions, and superior driveability. Ethanol and ETBE are among the oxygenates used in reformulated gasoline that is required in certain ozone non attainment areas in the United States.

8.1 Environmentally Friendly

Ethanol is most commonly used to power automobiles, though it may be used to power other vehicles, such as farm tractors and airplanes. Ethanol (E100) consumption in an engine is approximately 34% higher than that of gasoline (the energy per volume unit is 34% lower). However, higher compression ratios in an ethanol-only engine allow for increased power output and better fuel economy than would be obtained with the lower compression ratio. In general, ethanol-only engines are tuned to give slightly better power and torque output to gasoline powered engines. In flexible fuel vehicles, the lower compression ratio requires tunings that give the same output when using either gasoline or hydrated ethanol. For maximum use of ethanol's benefits, a much higher compression ratio should be used, which would render that engine unsuitable for gasoline usage. When ethanol fuel availability allows high compression ethanol-only vehicles to be practical, the fuel efficiency of such engines should be equal or greater than current gasoline engines. However, since the energy content (by volume) of ethanol fuel is less than gasoline, a larger volume of ethanol fuel would still be required to produce the same amount of energy.

A 2004 study, and paper published by the Society of Automotive Engineers, presents the possibility of a definite advance over hybrid electric cars' cost-efficiency by using a high-output turbocharger in combination with continuous dual fuel direct injection of pure alcohol and pure gasoline in any ratio up to 100% of either. Each fuel is stored separately, probably with a much smaller tank for alcohol, the peak cost efficiency being calculated to occur at approximately 30% alcohol mix, at maximum engine power. The estimated cost advantage is calculated at 4.6:1 return on the cost of alcohol used, in gasoline costs saved, when the alcohol is used primarily as an octane modifier and is otherwise conserved. With the cost of new equipment factored in the data gives a 3:1 improvement in payback over hybrid, and 4:1 over turbo-diesel (comparing consumer investment yield only). In addition, the danger of water absorption into pre-mixed gasoline and supply issues of multiple mix ratios can be addressed by this system.

All manufacturers recommend the use of ethanol for environmental reasons. A survey revealed that nine out of ten auto dealers use ethanol-blended gasoline in their personal vehicles. Over half of the dealerships surveyed indicated their customers reported benefits that included: reduced knocking and pinging, improved gas mileage, better acceleration, and improved starting qualities. A survey conducted at Iowa indicated that nine out of ten technicians used ethanol in their personal vehicles and reported the same benefits as the auto dealers. E-10 unleaded is approved under the warranties of all domestic and foreign automobile manufacturers marketing vehicles in the United States. The top three automakers in US, Daimler-Chrysler, Ford and General Motors, recommend the use of oxygenated fuels such as ethanol blends because of their clean air benefits and performance qualities. Ethanol is a good cleaning agent. It helps keep the engine clean in newer vehicles. In older vehicles it can sometimes loosen contaminants and residues that have already been deposited in a vehicle's fuel delivery system. Occasionally, these loosened materials collect in the fuel filter, and can then be removed simply by changing the fuel filter. All alcohols have the ability to absorb water. Condensation of water in the fuel system is absorbed and does not have the opportunity to collect and freeze. Since ethanol blends contain at least 10% ethanol, they are able to absorb water and eliminate the need for adding a gas-line antifreeze in winter. Ethanol is a fuel for old and new engine technology. Automotive engines older than 1969 with non-hardened valve seats may need a lead substitute added to gasoline or ethanol blends to prevent premature valve seat wear. Valve burning is decreased when ethanol blends are used because ethanol burns cooler than ordinary unleaded gasoline. Many high performance racing engines use pure alcohol for that reason.

Modern computerized vehicles perform better than non-computer equipped vehicles when operating correctly. The improvement in performance is due to the vehicle's computerized fuel system being able to make adjustments with changes in operating conditions or fuel type. Some of the chemicals used to manufacture gasoline, such as olefins, have been identified as a cause of deposits on port fuel injectors. Today's gasolines contain detergent additives that are designed to prevent fuel injector and valve deposits.

The quality of fuel used in any motor vehicle engine is very important to its long life and proper operation. Driveability will suffer if the fuel is not right for the air temperature or if fuel changes to a vapor incorrect. Gasoline is a complex mixture of approximately 300 various ingredients, mainly hydrocarbons, refined from crude petroleum oil for use as fuel in engines. Refiners must meet gasoline standards set by the American Society for Testing and Materials (ASTM), the Environmental Protection Agency (EPA), state regulatory agencies and their own company standards.

MTBE is a popular oxygenate that competes with ethanol for market share as an additive. It is the most widely used fuel additive to make reformulated gasoline, however it cannot be used as the dominant ingredient in a fuel. MTBE is highly corrosive, more volatile than ethanol, and more damaging to plastic and rubber fuel system components known as elastomers. Both MTBE and ETBE are high octane, low volatility, oxygenated fuel components made by combining alcohol with isobutlylene. MTBE is permitted in unleaded gasoline up to a level of 15% whereas ETBE can be added to gasoline up to a level of about 17%. Many car company warranties do not cover the use of methanol-based fuels, while all auto makers approve of the use of ethanol-blended gasoline. MTBE has been found to contaminate ground water and a ban on MTBE has been implemented in half the states in the United States with a reasonable possibility of the ban spreading to the remaining twenty five states in the near future. As long as a reformulated fuel mandate stays in place, ethanol is virtually guaranteed an expanded market. If the MTBE ban passes, then there is not enough ethanol production capacity to immediately meet excess demand caused by the ban. Therefore, more ethanol infrastructure would need to be built to accommodate the extra demand. A more gradual phase out would help the ethanol industry adapt to the growing need for its product.

8.1.1 Fuel Economy

All vehicles have a fuel economy (measured as miles per US gallon -MPG-, or liters per 100 km) that is directly proportional to energy content. Ethanol contains approx. 34% less energy per unit volume than gasoline, and therefore will result in a 34% reduction in miles per US gallon. For E10, the effect is small (~3%) when compared to conventional gasoline, and even smaller (1–2%) when compared to oxygenated and reformulated blends. However, for E85, the effect becomes significant. E85 will produce lower mileage than gasoline, and will require more frequent refueling. Actual performance may vary depending on the vehicle. This reduced fuel economy should be considered when making price comparisons. For example, if regular gasoline costs $3.00 per gallon, and E85 costs $2.19 per gallon, the prices are essentially equivalent. If the discount for E85 is less than 27%, it actually costs more per mile to use.

Research is underway to increase fuel efficiency by optimizing engines for ethanol-based fuels. Ethanol's higher octane allows an increase of an engine's compression ratio for increased thermal efficiency. In one study, complex engine controls and increased exhaust gas recirculation allowed a compression ratio of 19.5 with fuels ranging from neat ethanol to E50. Thermal efficiency up to approximately that for a diesel was achieved. This would result in the miles per gallon of a dedicated ethanol vehicle to be about the same as one burning gasoline. There are currently no commercially-available vehicles that make significant use of ethanol-optimizing technologies, but this may change in the future.

While the energy content of ethanol is approximately 33% less than gasoline the difference can be partially offset by improved thermal efficiency (Mueller et al. 2018; Stein et al. 2013). The authors state that increased ethanol enables redesigned engines to operate at higher compressions ratios. The study cites Ford's Ecoboost direct injection engine tests that showed that 96 RON E20 at 11 0.9: 1 CR provides comparable fuel economy. Stein restates that volumetric fuel economy can stay equal to gasoline for E20-E30 based on several efficiency effects including reduced enrichment with higher ethanol content, and improved efficiency at part loads due to reduced heat transfer losses with ethanol, as well as the above mentioned higher compression ratios. In 2016 Oak Ridge National Laboratory conducted engine tests on different ethanol blends to demonstrate the fuel economy of different ethanol blends in dedicated engines with downsizing and down speeding (O. R. N. Laboratory, https://info.ornl.gov/sites/publications/files/pub61169.pdf, ORNL/TM2016/42). Down speeding was achieved with larger drive wheels and a different differential. Downsizing was achieved with increased test weight. For E30 (101 RON) the results showed already a fuel economy gain of 5% for the unmodified vehicles and a fuel economy improvement of 10% for the modified (downsped/downsized vehicle) over the baseline E10. Furthermore, the results showed that a splash blended RON 97 with 15% ethanol already in an unmodified 2014 Ford Fiesta (non-FFV) vehicle with a small turbocharged direct-injection engine already showed quasi fuel economy parity for the US06 driving cycle. Also noteworthy is that these tests do not include further potential improvements from custom designed pistons to increase the compression ratio. These recent research findings show that the lower energy density of ethanol will likely not be a significant detriment to fuel economy in properly designed fuels and modern engines and may even be a an advantage in future high octane dedicated engine designs. In iBEAM all emissions calculations revert to a per distance driven basis and are therefore independent of fuel economy.

References

Al-Mashhadani H, Fernando S (2017) Properties, performance, and applications of biofuel blends: a review. AIMS Energy 5:735–767

Bajpai P (2007) Bioethanol. PIRA Technology Report, Smithers PIRA, UK

Chrysler Corporation (1997) Chrysler corporation's flexible fuel minivans

Costa RC, Sodre JR (2010) Hydrous ethanol vs. gasoline-ethanol blend: engine performance and emissions. Fuel **89**(2010):287–293

Fivgal A, Speranza LG , Branco CM, Ouadi M and Hornung A (2019). A review on the current state of the art for the production of advanced liquid biofuels. AIMS Energy **7**(1):46–76

Ford Motor Company (1998) The 1998 E85 ford taurus flexible fuel vehicle

Mueller S, Unnasch S, Keesom B, Mohan S, Goyal L (2018) The impact of higher ethanol blend levels on vehicle emissions in five global cities. University of Illinois at Chicago Energy Resources Center. https://www.erc.uic.edu/assets/pdf/UIC5cities_FINAL.pdf

Niven RK (2005) Ethanol in gasoline: environmental impacts and sustainability review article. Renew Sustain Energy Rev 9:535–555

O. R. N. Laboratory, Summary of High-Octane, Mid-Level Ethanol Blends Study. https://info.ornl.gov/sites/publications/files/pub61169.pdf, ORNL/TM2016/42

Owen K, Coley T (1995) Automotive fuels reference book, 2nd edn. Society of Automotive Engineers, Inc., USA

Rasskazchikova TV, Kapustin VM, Karpov SA (2004) Ethanol as high-octane additive to automotive gasolines. Production and use in Russia and abroad. Chem Technol Fuels Oils **40**(4):203–10

RFA—Renewable Fuels Association (1999) Ford coupon program to encourage use of E85. Ethanol report

RFA—Renewable Fuels Association (2001) Ethanol: clean air, clean water, clean fuel—industry outlook 2001. https://www.ethanolrfa.org/RFAannualreport01.pdf

RFA—Renewable Fuels Association (2006a) Ethanol industry outlook: from niche to nation. www.ethanolrfa.org/objects/pdf/outlook/outlook_2006.pdf

RFA—Renewable Fuels Association (2006b) Fuel ethanol industry plants and production capacity

RFA—Renewable Fuels Association (2006c) Statistics data. https://www.ethanolrfa.org/industry/statistics/#C

RFA—Renewable Fuels Association (2007a) Ethanol industry outlook: building new horizons. https://www.ethanolrfa.org/objects/pdf/outlook/RFA_Outlook_2007.pdf

RFA—Renewable Fuels Association (2007b) Cellulosic ethanol grants provide much needed boost to fledgling technology. www.biofuelsjournal.com/articles/RFA

Silva R, Cataluña R, Menezes EW, Samios D (2005) Piatnicki CMS (2005) Effect of additives on the antiknock properties and Reid vapor pressure of gasoline. Fuel 84:951–959

Stein R, Anderson J and Wallington T (2013) An overview of the effects of ethanolgasoline blends on SI engine performance, fuel efficiency, and emissions. SAE Int J Eng **6**(1):470–487, 013

Chapter 9
Benefits and Problems with Bioethanol

Abstract Benefits and Problems with ethanol/ ethanol blends are discussed in this chapter. Ethanol is easily biodegraded in the environment, and produces much less pollutants in internal combustion engines than petroleum fuels. Ethanol/ethanol blends create problem during storage, transportation and combustion.

Keywords Biofuels · Bioethanol · Biodegradable · Internal combustion engine

9.1 Benefits

Ethanol provides several benefits. It is easily biodegraded in the environment, and produces much less pollutants in internal combustion engines than petroleum fuels (Bajpai 2007; Green fuels 1998a, b). It has low toxicity and is miscible with water. Many car makers are producing more vehicles with tolerances to burn high %E fuels more efficiently. Thus, the risk posed by ethanol to the environment is significantly lower than that of fuels produced from petroleum and the demand for ethanol will increase with time as these automobile improvements take place.

Ethanol is miscible with gasoline in any proportion, but is found most commonly as 10% ethanol and 85% ethanol. FFVs can operate on blends of ethanol and gasoline anywhere between 0 and 85% ethanol. Benefits of ethanol include: higher performance; cleaner burning fuel; positive energy balance; currently cheaper than gasoline (after considering subsidies). A notable fact is that fuel ethanol is versatile and can basically be used in two capacities: a fuel additive (E10) and an almost stand-alone fuel (E85). Different issues surround each type of fuel. Any internal combustion engine, including small engines such as those in lawn mowers, can use a blend up to 10% ethanol. Ethanol has a higher octane rating than most gasoline, which means that engines burning ethanol are less likely to "knock." Pure ethanol has an octane number of 112 while E85 is about 105 octane. Such a high octane rating means that even high performance engines can use ethanol fuels. Ethanol is a good racing fuel because it combusts at a lower temperature than gasoline, thus requiring less cooling power from the radiator.

P. Bajpai, *Developments in Bioethanol*, Green Energy and Technology,
https://doi.org/10.1007/978-981-15-8779-5_9

Ethanol as a fuel has a positive energy balance, which means that combustion of one unit of ethanol yields more energy than the aggregate energy required to produce the inputs that make up that unit. It follows that ethanol is more beneficial for the environment than using only petroleum based fuels (Table 9.1).

The United States. Environmental Protection Agency considers ozone to be the most common air pollution problem (EPA 1997). To combat this problem, ethanol is widely used in reformulated gasolines to help urban cities meet public health standards for ozone. Because it's produced from renewable resources, ethanol is the only transportation fuel that reduces greenhouse gas emissions from cars. Fossil fuels release carbon trapped in the soil into the air, where it reacts with oxygen to form carbon dioxide, a greenhouse gas that traps the earth's heat, contributing to global warming (EPA 2001).

The United States government is supporting production of ethanol with methods using lesser energy in comparison to traditional fermentation and that use cellulosic biomass, which needs less cultivation, fertilizer, and pesticides as compared to corn or sugar cane. Cellulosic ethanol raw materials includes native prairie grasses, fast growing trees, sawdust, and even waste paper.

Table 9.1 Benefits of Bioethanol

Ethanol is a renewable energy source as it results from the conversion of the sun's energy into usable energy. Ethanol formation starts with photosynthesis. This causes feedstocks to grow which are converted into ethanol
Exhaust gases from ethanol burn more cleanly
Ethanol-blended fuel as E10 which contains 10% ethanol and 90% gasoline reduces greenhouse gases to the extent of 3.9%
The use of ethanol blended fuels such as E85 which contains 85% ethanol and 15% petrol reduces the net emissions of greenhouse gases to the extent of 37.1%, which is quite substantial
The energy balance is positive. Depending on the type of raw material it varies from 1.24 to 8. During the production, the output of energy is more than the input
Any plant containing sugar and starch can be used for producing ethanol can be used. Sugar cane, barley, wheat, potatoes etc. can be used
The carbon dioxide produced in the bioethanol process is the same quantity as the one the crops earlier absorbed during photosynthesis
Use of ethanol reduces ozone formation which is an significant environmental issue
The emissions released during burning of ethanol are less reactive with sunlight in comparison to those produced by burning petrol, which results in a reduced potential for producing the ozone
Energy security is benefitted. Ethanol shifts the requirement for some foreign produced oil to locally produced energy sources. Countries not having any access to crude oil resources can grow crops for energy use and get some economic freedom
The addition of high octane additives is reduced
The fuel spills get easily biodegraded or diluted to non toxic levels

https://bioethanol-np.blogspot.com/p/advantages-of-bioethanol.html

Ethanol is made from agricultural crops, which take carbon dioxide and give off oxygen. This maintains the balance of carbon dioxide in the atmosphere. Increased use of renewable fuels like ethanol, will help counter the pollution and global warming effects of burning gasoline. Under current conditions, use of ethanol blended fuels as E85 can reduce the net emissions of greenhouse gases by as much as 30–36% and can further contribute by decreasing fossil energy use by 42–48% (Hu et al. 2004). Ethanol blended fuel as E10 reduces greenhouse gases by 2.4–2.9% and fossil energy use by 3.3–3.9%.

9.1.1 Environmental Friendliness

The search for reduced dependency on oil producing countries and the reduced net ouptut of the greenhouse gas carbon dioxide are some of the arguments in favor of ethanol. According to some opponents it is mainly a government subsidy for corn growing agricultural business. The Archer Daniels Midland Corporation of Decatur, Illinois, (ADM) is the largest grain processor in the world. It is producing 40% of the ethanol used for producing gasohol in the United States. The company has been persuasive in their defense of ethanol and liberal in contributing to both political parties.

According to other opponents "it is economically absurd to consider ethanol from grain as a replacement for petroleum, when industrial ethanol is made from petroleum feedstocks because it is far cheaper than fermented ethanol. Environmentalists do not like arguments like this, since they advocate a transition to renewable energy rather than continued usage of fossil fuels.

There is widespread belief that ethanol containing fuel is more environmentally friendly than gasoline without additives. However, there is a controversy over whether requiring ethanol in automotive fuel is wise since it has been argued that the beneficial effects of ethanol can be achieved with other cheaper additives made from petroleum. Also, both the Environmental Protection Agency and the National Academy of Sciences have issued reports showing that adding ethanol to gasoline will at best have no effect on air quality and could even make it worse. Studies show ethanol could even increase emissions of nitrogen oxides and volatile organic compounds, which are major ingredients of smog" (www.worldofmolecules.com).

9.2 Problems with Ethanol/Ethanol Blends

Table 9.2 presents problems with ethanol/ethanol blends create problem during storage, transportation and combustion (Orbital Engine Company 2002a, b, 2003). Main drawback of using ethanol with gasoline is phase separation because it is immiscible with diesel fuel over a wide range of temperatures.

Table 9.2 Problems and Concerns of bioethanol

The amount of cultivable land needed for growing the crops for producing a large amount of fuel is enormous. This has a great impact on the biodiversity of our environment as we are seeing natural habitats being overrun, including forests
Bioethanol is neutral. There is an argument over this issue when all elements are taken into consideration including the cost of changing the land use of an area, transportation and the burning of the crop
When bioethanol is produced, a large amount of carbon dioxide is released making its environmental usefulness closer to zero
When ethanol is produced, the air is filled with greenhouse gases in the amounts which can be compared with the emissions of internal combustion engines
The energy content of bioethanol is 70% of that of gasoline. So it is less efficient in comparison to petroleum. One has to pay more for more fuel for doing the same work. Therefore, the use of bioethanol in public transportation would increase the price of the service
Many older cars are not presently equipped for handling even 10% ethanol whereas there is concern that use of 100% ethanol reduces fuel economy by about 15 to 30% as compared to 100% petroleum
Bioethanol has an octane number of around 105. It can be burnt in the engines having much higher compression ratio. The engines designed for working on the new energy cannot be used for their petrol variants
Harmful effect on electric fuel pumps. Increases internal wear and unwanted spark generation
Not well-suited with capacitance fuel level gauging indicators and could show wrong fuel quantity indications in vehicles using that system
Phosphorous and nitrogen used in the production of ethanol have deleterious effect on the environment
It absorbs water from the air and therefore has high corrosion destructiveness. Therefore, it is transported only by auto transport or railroad
With pure ethanol, starting a car in cold weather is difficult because it is difficult to vaporise so most fuels retain a small amount of petrol—such as E85 cars with 85% ethanol and 15% petroleum
A large amount of cultivable land is needed for growing the crops. This could see destruction of some natural habitats including rainforests
Because of the profitable prices of bioethanol some farmers may sacrifice food crops for bioethanol production which will increase food prices around the world
Ethanol also absorbs water easily; corrodes the materials
Large infrastructure change is needed for providing ethanol refueling station
Use of higher amount of ethanol reduces fuel economy

https://sites.lafayette.edu/egrs352-sp15-biofuels/drawbacks-of-bioethanol/bioethanol-np.blogspot.com-advantages-of-bioethanol

One of the most challenging issues related to ethanol fuel blends involves the stability of mixtures. The shelf life of ethanol fuel blends is much lower due to its water absorbing and corrosive qualities. It does not store longer than 2 or 3 months without adding a stabilizer (US DOE 2006a, b). Even so, Ethanol stored in fiberglass or plastic tanks will make the fiberglass soft and mushy, leading to tank failure and engine failure, because the plastic or fiberglass is dissolving into the gas.

Ethanol can't travel in pipelines along with gasoline, because it picks up excess water and impurities. As a result, ethanol needs to be transported by trucks, trains, or barges, which is more expensive and complicated than sending it down a pipeline. It would be a much expensive affair to build a completely new pipeline network specifically for ethanol.

Some of the soft metals such as zinc, brass, lead, aluminium, copper are not compatible with ethanol and can suffer corrosion and pitting if exposed to it for extended periods (US DOE 2006b). Furthermore, ethanol dissolves in water and is more electrically conductive than petrol. The presence of water can facilitate corrosion and the conductivity facilitates the possibility of galvanic corrosion.

Ethanol has a solvent effect and will loosen gums and other deposits in fuel systems that have been operated on mineral petrol a long time. In extreme cases this can clog fuel filters and cause the engine to run poorly. Degradation of fuel system components can potentially occur because the materials used for hoses, seals, O-rings, membranes and gaskets are not compatible with ethanol. The typical mechanism is that the ethanol is absorbed into the material and breaks down the molecular bonds within it. This can lead to swelling of the material, softening or embitterment and eventually failure of the component.

Ethanol blend fuels have an affinity to absorb much greater extent of water, very quickly, compared to conventional non-alcohol gasoline (ethanol gasoline blends can absorb 50 times more water than conventional non alcoholic gasoline). Some water can be dissolved in ethanol-petrol blends and will pass through the fuel system with no effect. However, if the amount of water present is too great, the blend will separate into an upper petrol layer and a lower water/ethanol layer. Generally fuel is drawn from the bottom of the tank so the water/ethanol layer will be drawn into the engine first and the engine will not run (Lapuerta et al. 2007; Bhattacharya and Mishra 2003).

The ethanol molecule contains oxygen while the main components of petrol do not. The effect of this is that less oxygen from air is required to achieve complete combustion and so if the air/fuel mixture is not adjusted, the mixture is leaner than it would be on pure petrol. This can lead to engine operability problems such as hesitancy at full throttle and/or higher exhaust temperatures (Orbital Engine Company 2002a). The volatility characteristics of an ethanol/petrol blend differ from those of the unblended petrol on its own. These differences in volatility can impact on engine operability.

Ethanol fuel blend is not compatible with some vehicle parts like capacitance fuel level gauging indicators as they give erroneous fuel quantity indications in vehicles. The use of ethanol-based fuels can negatively affect electric fuel pumps by increasing internal wear and undesirable spark generation.

Small scale production of ethanol requires a significant input of equipment and labour. There is added cost to process the alcohol to the required 200 proof for blending with gasoline for unmodified engines. To be most effective, the ethanol should be used in modified engines.

An Ethanol and gasoline mixture can cause damage to the environment during transportation, storage and consumption. Once the mixture is leaked or spilled into surface or ground water, ethanol will hamper degradation of other toxic components in gasoline. Ethanol also increases water solubility of gasoline, leading to larger contamination areas. Moreover, improvement of air quality is still uncertain. While an ethanol and gasoline mixture has lower sulphuric oxide and carbon dioxide emissions, it has higher nitrogen oxide and VOC emissions, which causes ozone depletion (Jia et al. 2005).

References

Bajpai P (2007) Bioethanol. PIRA Technology Report, Smithers PIRA, UK

Bhattacharya TK, Mishra TN (2003) Studies on feasibility of using lower proof ethanol-diesel blends as fuel for compression Ignition engines. Inst Eng Agric Eng 84:56–59

Department of Energy (2006a) Bioethanol feedstocks. https://www1.eere.energy.gov/biomass/abcs_biofuels.html

Department of Energy (2006b) Guidebook for handling, storing, & dispensing fuel ethanol prepared for the U.S. department of energy by the center for transportation research energy systems division, Argonne National Laboratory

EPA (Environmental Protection Agency) (1997) Fact sheet: EPA's revised ozone standard, United States. www.epa.gov/ttn/oarpg/naaqsfin/o3fact.html. Accessed Mar 2001

EPA (Environmental Protection Agency) (2001) Fact sheet: global warming: climate, United States. https://www.epa.gov/globalwarming/climate/index.html

Green Fuels (1998a) Environmental effects of ethanol and gasoline. https://www.greenfuels.org/ethaenv1.html

Green Fuels (1998b) How does ethanol clear the air? https://www.greenfuels.org/ethaair.html

Hu Z, Pu G, Fang F, Wang C (2004) Economics, environment, and energy life cycle assessment of automobiles fueled by bio-ethanol blends in China. Renew Energy 29:2183–2192

Jia LW, Shen MQ, Wang J, Lin MQ (2005) Influence of ethanol–gasoline blended fuel on emission characteristics from a four-stroke motorcycle engine. J Hazard Mater 123(1–3):29–34

Lapuerta M, Armas O, García R (2007) Stability of diesel–bioethanol blends for use in diesel engines. Fuel 86(10–11):1351–1357

Orbital Engine Company (2002a) A literature review based assessment on the impacts of a 10 and 20 % ethanol gasoline fuel blend on the Australian vehicle fleet. Canberra

Orbital Engine Company (2002b) A literature review based assessment on the impacts of a 20 % ethanol gasoline fuel blend on the Australian vehicle fleet. Canberra

Orbital Engine Company (2003) Market barriers to the uptake of a biofuels study. A testing based assessment to determine impacts of a 20% ethanol gasoline fuel blend on the Australian passenger vehicle fleet-2,000 hrs material compatibility testing. Canberra, https://sites.lafayette.edu/egrs352-sp15-biofuels/drawbacks-of-bioethanol/, www.worldofmolecules.com, https://bioethanol-np.blogspot.com/p/advantages-of-bioethanol.html

Chapter 10
Global Production of Bioethanol

Abstract Several countries have introduced or are introducing programmes for fuel ethanol. Ethanol production in different parts of world is discussed. Brazil and the United States are the dominant industrial players, accounting for 87% of global biofuel production driven by government support.

Keywords Biofuels · Bioethanol · Fuel ethanol · Global biofuel production

According to the International Energy Agency (IEA) production of biofuels is expected to increase by 15% during the next five years, reaching 165 billion liters. It is predicted that by 2023, biofuels will account for about 90% of the renewables used in transport. Fuel ethanol accounts for about two third of biofuel production growth, whereas biodiesel and hydrotreated vegetable oil account for the rest (http://ethanolproducer.com/articles/15673/iea-predicts-growth-in-global-ethanol-production-through-2023).

The countries in Asia account for most of the growth in biofuel output during the next five years. China, India and member states of the Association of Southeast Asian Nations represent half of the worldwide expansion in biofuel production. Latin America is responsible for an extra 45% of that growth, especially Brazil.

"Under favorable market and policy conditions, the IEA said the growth in biofuels could be even more significant over the next five years, reaching nearly 206 billion liters of production. According to the IEA, ethanol production increased by 3% globally last year, reaching 104 billion liters of production. Increased production was primarily realized in the United States, European Union and China. Ethanol production is currently expected to expand at an average annual growth rate of 2%, reaching 119 billion liters by 2023" (http://ethanolproducer.com/articles/15673/iea-predicts-growth-in-global-ethanol-production-through-2023).

Growth in ethanol production is driven by Brazil, followed by China, expanding markets in India and Thailand. IEA predicts that China's share of global production will increase in the next few years but the United States and Brazil will still account for 80% of worldwide output (RFA 2018).

P. Bajpai, *Developments in Bioethanol*, Green Energy and Technology,
https://doi.org/10.1007/978-981-15-8779-5_10

Table 10.1 Fuel ethanol production worldwide in 2019, by country (in million gallons)

Region	2014	2015	2016	2017	2018	2019
United States	14,313	14,807	15,413	15,936	16,061	15,800
Brazil	6,760	7,200	6,760	6,860	7,920	8,620
European Union	1445	1387	1377	1,400	1,430	1,440
China	635	813	845	860	1,050	900
Canada	510	436	436	470	480	500
India	85	195	275	210	400	530
Thailand	310	334	322	370	390	420
Argentina	160	211	264	290	290	290
Rest of World	865	391	490	414	549	600
Total	25,083	25,774	26,182	26,810	28,570	29,100

https://ethanolrfa.org/statistics/annual-ethanol-production/

Under favorable conditions, ethanol production could increase by as much as 26 billion liters, reaching 145 billion liters 2023, with growth primarily in the United States, Brazil, China, India and Argentina. This level is 22% higher than the IEA's main case forecasts.

According to IEA, annual production of novel advanced biofuels will reach 1.4 billion liters by 2023. This shows a three times increase from 2017. Cellulosic ethanol accounts for 60% of production within the novel advanced biofuels, with the rest from advanced biofuels for the diesel pool and aviation biofuels (https://ethanolproducer.com%3Earticles%3Eiea%2Dpredicts%2Dgrowth%2Din).

"An accelerated case for novel biofuels suggest nearly 2.3 billion liters could be produced by 2023 if the right conditions are present. Under this scenario, the share of cellulosic ethanol would increase to two-thirds of production, with one-third from advanced biofuels used within the diesel pool and aviation biofuels" (https://ethanolproducer.com%3Earticles%3Eiea%2Dpredicts%2Dgrowth%2Din).

Several countries have introduced or are introducing programmes for fuel ethanol (Johnson et al. 2010; Yan et al. 2010; Zhenhong 2006; Piacente 2006. Bajpai 2007; Goldemberg 2008). According to RFA, Brazil and the United States are the dominant industrial players, accounting for 87% of global biofuel production driven by government support (https://www.statista.com/statistics/281606/ethanol-production-in-selected-countries/;;https://ethanolrfa.org/statistics/annual-ethanol-production/). This statistic shows the fuel ethanol production in major countries and regions in 2019. In that year, the United States produced the highest amount of ethanol in the world, producing some 15.8 billion gallons in total. With almost 8.6 billion gallons, Brazil was ranked second. Ethanol is a grain alcohol which can be mixed with gasoline and used in motor vehicles. Table 10.1 shows fuel ethanol production in the world in 2019, by country (in million gallons). Worldwide production of ethanol based by country in percent is shown in Table 10.2. The United States is the largest producer of ethanol. Brazil is the next biggest producer.

Below is a description about bioethanol programs worldwide:

Table 10.2 Fuel ethanol production—% of World Production

	% of World Production (%)
United States	54
Brazil	30
European Union	5
China	3
Canada	2
India	2
Thailand	1
Argentina	1
Rest of World	2

10.1 European Union

The European Commission is supporting biofuels with the following objectives:

- Reduction of greenhouse gas emissions
- Boosting the decarbonisation of transport fuels
- Diversifying fuel supply sources
- Offering new income opportunities in rural areas
- Developing long-term replacements for fossil fuel.

In European Union, Ethanol is mainly produced from sugar beets and wheat. Sugar beets prove to be a good feedstock for European bioethanol production. Because sugar beets have a much larger yield per hectare than wheat, the European Union currently produces 2 million more tons of sugar beet than wheat on 20 million less hectares of land. Additionally, sugar beets produce more ethanol per hectare: a hectare of sugar beets can produce 30 hectoliters more ethanol, on average, than wheat. Also, sugar beet ethanol is shown to have a more energy-efficient production process than wheat ethanol. Currently, the most important bioethanol producers are France, Spain, Germany, Sweden, Poland and Italy. In comparison with the United States and Brazil, European Union ethanol for fuel production is still very modest. These 2 countries are now in competition for the title of world biggest producer. For both these 2 countries the main driver is reducing dependency from foreign fossil fuel sources.

In Europe, except for Sweden, and unlike the United States or Brazil, ethanol is not incorporated directly but transformed into ETBE (obtained by reacting isobutene, a liquefied petroleum gas, with ethanol) before being blended with gasoline. One reason for this regional particularity is the obligation to properly account for motor fuel properties such as volatility, since pure ethanol makes ethanol/gasoline blends more volatile. Another advantage of this practice is that it avoids separation of the alcohol and "gasoline" phases in the presence of traces of water.

USDA Foreign Agricultural Service's Global Agricultural Information Network discusses the present use of biofuels in the European Union and the potential for greater use in the future

"EU has set a 10% target for renewable energy use in transport for 2020. That target increases to 14% in 2030, with advanced biofuels counting double to the target. Taking double-counting into account, biofuels accounted for 7.1% of energy use in transportation last year and are expected to increase to 7.3% this year. The increase is expected to be supported by increased imports" (https://www.fas.usda.gov/data/eu-28-biofuels-annual-1).

Fuel ethanol production reached about 5.443 billion liters in the European Union in 2018, which is higher from 5.38 billion liters in 2017. European Union fuel ethanol production increased to 5.505 billion liters in 2019.

Fuel ethanol imports reached about 505 million liters in 2018, up from 238 million liters in 2017. Fuel ethanol imports are increased slightly in 2019, reaching 570 million liters.

The European Union exported about 91 million liters of fuel ethanol last year, up from 41 million liters in 2017. Exports this year are expected to reach 95 million liters. Ethanol consumption has reached 7.22 billion liters in 2019, up from 7.082 billion liters in 2018 and 6.89 billion liters in 2017.

The average ethanol blend rate in the European Union was 5.7% in 2018, up from 5.4% in 2017. The blend rate is expected to increase to 5.9% this year.

European Union is presently having 58 first generation ethanol plants. This number has remained stable since 2017. The capacity of those facilities was 8.66 billion liters in 2018, up from 8.6 billion liters in 2014. Capacity remained at 8.66 billion liters through 2019. Capacity use is expected to be higher slightly, from 70% in 2017 and 2018 to 71% in 2019. The European Union also presently has two cellulosic ethanol refineries, having a joint capacity of 60 million liters.

"Sugar beet is expected to be the primarily feedstock for European Union ethanol production this year, reaching 8.145 million metric tons, followed by wheat with 5.665 million tons, corn with 5 million tons, triticale with 720,000 tons, barley with 430,000 tons and rye with 418,000 tons. Germany is expected to be the top European Union consumer of ethanol this year, with 1.505 billion liters, followed by the U.K. with 925 million liters and France with 880 million liters. Spain, Poland, the Netherlands, Italy and Sweden are also among the top eight ethanol consuming countries in the European Union" (http://ethanolproducer.com/articles/16422/report-eu-ethanol-consumption-to-increase-in-2019).

France will become the major ethanol producing country in the European Union. It is producing "1 billion liters, followed by Germany at 785 million liters, the United Kingdom with 695 million liters, Hungary with 645 million liters, Belgium with 645 million liters, the Netherlands with 565 million liters, Spain with 522 million liters, Poland with 265 million liters and Austria with 235 million liters" (ethanolproducer.com).

The German Bioethanol Industry Association has published the 2018 data for the production and consumption of bioethanol. German bioethanol production, standardised for use as fuel, dropped to 613 000 tonnes, a reduction of 8.9% in comparison with the preceding year. But, in a shrinking petrol market, sales of bioethanol increased by about 3.0% in 2018, when the fuel market crashed overall, with 17.8 million tonnes of petrol sold—about 2.5% less petrol in comparison to 2017: (18.3 million tones). The consumption of bioethanol, which is used as a petrol admixture for the petrol types Super E10, Super Plus and Super (E5) or used to produce ETBE (ethyl tert-butyl ether), increased to just below 1.2 million tones (bioenergyinternational.com).

Whereas just under 110 000 tonnes of bioethanol were used for ETBE production, about 1.4% less than in the earlier year, the admixture percentage increased drastically. This increased the percentages of bioethanol in the petrol types Super E5, Super E10 and Super Plus.

Stefan Walter, Managing Director of the BDBe commented on the official mineral oil data, explained that the obligation for reducing greenhouse gas emissions for all fuels from the present level of 4.0% is "clearly too low to ensure even more climate change mitigation in the transport sector".

But, this has changed in 2020, when the greenhouse gas savings quota has increased to 6% as stipulated by law. The mineral oil industry will use the biofuels which meet the commitments.

The debate about meeting climate change mitigation targets, about the necessary measures like high price for fossil based fuels, the promotion of electromobility and the use of alternative fuels are also, in the view of BDBe, increasing public awareness of the potential biofuels have to reduce carbon dioxide emissions and also improve the quality of urban air.

The use of bioethanol in petrol is already reducing carbon dioxide emissions in the transport sector by 3.1 million tonnes. In mathematical terms, this is equivalent to about 1.0 million passenger cars without carbon dioxide emissions. Comparative tests started by the BDBe which use the new WLTP measurement method show that the use of Super E10 not only substantially reduces carbon dioxide emissions but also the particulate matter and nitrogen oxides (https://bioenergyinternational.com/markets-finance/sales-of-bioethanol-increased-in-germany-in-2018-despite-drop-in-gasoline-consumption).

10.2 Australia

The Australian government has been supporting ethanol since 2000 with a range of tax exemptions and production subsidies.

Production capacity of bioethanol in Australia is presently 440 ML per year (0.48% of global bioethanol production). Its capacity for biodiesel production is 100 ML per year (0.17% of the global output). "The annual production of these two biofuels have an equivalent amount of energy as 375 ML of diesel. Currently, domestic biodiesel production is approximately one per cent of total domestic diesel consumption, indicating biofuel's extremely small market share. In addition, there is currently no aviation biofuel production in Australia" (https://arena.gov.au/assets/2019/11/biofuels-and-transport-an-australian-opportunity.pdf).

The Australian car fleet can use substantial volumes of biofuel. In 2015, 95% of the 2.5 million cars in Queensland alone were compatible with ethanol. This figure has reached 98% in 2020. Most automakers are now manufacturing cars fit for higher ethanol fuel blends. Australian plants can presently only produce sufficient biofuel to satisfy 0.6% of the nation's demand for transport fuel, which is very much behind the global average of 2.7%. This capacity has reduced in the last 10 to 15 years because of plant closures. In the meantime, existing plants are working below the full capacity.

In Brazil car manufacturers have also shown the ability of industry for developing engines capable of running on high biofuel blends. Brazil now has more than 25 million 'flex' vehicles, which can run on 0 to 100% ethanol. There is a increasing domestic awareness of Australia's lagging biofuel production relative to the global average, and of the advantages and opportunities higher production would bring to the broader community.

"This awareness is resulting in state government initiatives like the Advance Queensland Biofutures 10-Year Roadmap and Action Plan, and the Bioenergy Roadmap for South Australia. A 40-fold expansion over the next 30 years is a significant challenge for the industry. Achieving it will require sustainable growth in feedstock resources, the application of new technologies and a consistent development of new production facilities. The investment required by production facilities alone is estimated at between $25 billion and $30 billion" (www.cefc.com.au).

The Australian biofuel industry is although small, it has immense experience using low value waste streams for remaining viable in periods of policy uncertainty and low consumer demand. But it is hindered by higher costs, juvenile technology, its small scale. Also a suitable policy and regulatory framework is lacking.

Table 10.3 shows the biofuel plants in Australia, including production capacity. Especially, several biodiesel plants built in the early 2000s are not operating anymore. The shutting down of these plants is the result of many factors. These include reduced cost imports monopolising the Cleaner Fuels Grant Scheme, increasing the price of raw material and inconsistent product quality.

Table 10.3 Operating commercial-scale bioethanol plants in Australia

Company	Location	FEEDSTOCK	PRODUCTION
Manildra Group	Bomaderry, NSW	Waste starch	300 ML/yr bioethanol
United Petroleum	Dalby, Qld	Red sorghum	80 ML/yr bioethanol
Wilmar Sucrogen	Sarina, Qld	Sugarcane	60 ML/yr bioethanol

CEFC and ARENA (2019). Biofuels and Transport: An Australian opportunity https://arena.gov.au/assets/2019/11/biofuels-and-transport-an-australian-opportunity.pdf

Australia has larger areas of cultivable and semi-cultivable land; established agricultural, forestry and engineering industries; important raw materials and superb solar resources. All of these give Australia a exclusive opportunity for meeting its projected biofuel requirements.

In Australia, bioethanol is produced from grain and sugar cane molasses. Australia only produces small amounts of biodiesel to date. However, several industrial scale plants are on the drawing board or in the construction phase. The fuel ethanol industry is still waiting for the big breakthrough, but political support for the industry is growing. Australia is a net importer of crude oil, and biofuels are considered to be increasingly important in view of surging world oil prices. Advocates of ethanol production also cite benefits for the ailing domestic sugar cane industry.

10.3 China

China is the world's fourth largest fuel ethanol producer and consumer after the United States, Brazil, and the European Union. China has ambitious future growth targets for bioethanol from second generation waste biomass (Liu 2006; Zhenhong 2006; Yan et al. 2010). In China bioethanol plants use corn, wheat and cassava. Sweet sorghum and sugar cane have future potential. "Regarding second generation feedstocks, in 2008, COFCO built the first non-grain, 2million tons cassava fuel ethanol factory in Guangxi. In 2012, Henan Tianguanbuilt built its first 10 k-ton cellulosic fuel ethanol pilot program. In 2013, Shandong Longlive built its 50 k-ton corn cob fuel ethanol production line. In 2014, Zonergy built its 30 k-ton sweet sorghum fuel ethanol factory" (Guoqing 2018; noaw2020.eu).

Ethanol gasoline is now being promoted all through China. Fuel ethanol production will be increased by 5 to 6 times on the base of present scale. A cellulosic fuel ethanol plant is being constructed at the capacity of 50,000 ton/y. By 2025—attempts will be made to realize large scale production of cellulosic fuel ethanol. "The overall development of advanced biological liquid fuel technologies, equipment and the industry will reach an international advanced level, and an improved market oriented operating mechanism will be established" (noaw2020.eu›2018/12›C1-1-COFCO-Wuguoqing).

So far, Chinese government has made mandatory the use of E10 in nine provinces (at central and northern China) that account for about one-sixth of that country's vehicles. Officials say that this mandate aims at reducing the oil demand—that is currently 40% imported—and also aims at improving air quality in big cities. However, Chinese authorities also frequently highlight the targets of helping stabilize grain prices and raising farmer's income.

In China, more than 80% of ethanol is produce from grains (corn, cassava, rice etc.), about 10% from sugar cane, 6% from paper pulp waste residue and the remainder is produced synthetically. So far, no significant amounts of biodiesel have been produced in China. There are no industrial scale biodiesel plants in the country. Fuel ethanol is exempted from consumption tax (5%) and value-added tax (17%).

In the recent years, the fuel ethanol market in China has remained narrow minded. Imports were not allowed till 2015, and China hardly ever produced surplus volumes to export. As additional duties were implemented earlier this year, in China, fuel ethanol market backed away further from the global market. China is producing a wide range of ethanol products on a commercial-scale, including fuel ethanol, potable alcohol and industrial chemicals. Contrasting other major ethanol producers, China's major end use market for ethanol is to produce industrial chemicals and not fuel ethanol. In the past, China's ethanol output has followed national policy priorities. Since 2016, China's corn processors, including fuel ethanol and industrial chemical producers, are enjoying the advantage of corn processing subsidies based on throughput volumes. In addition, China is expanding blending of ethanol with gasoline on a nationwide basis by doubling the number of administrative regions implementing E10, and investments supported by government for expanding the production capacity. Inspite of strong central government support through policies and financial backing, ethanol sector of China faces near-term structural challenges and long term feedstock supply challenges for producing enough fuel ethanol for meeting ambitious E10 goals. Most experts consider China's E10 consumption target to be virtually unachievable at the existing pace of market development (https://apps.fas.usda.gov/newgainapi/api/report/downloadreportbyfilename?filename=Biofuels%20Annual_Beijing_China%20-%20Peoples%20Republic%20of_8-9-2019.pdf).

In 2019, fuel ethanol consumption was estimated at a record 4,311 million liters (3.4 million tons), up 1,397 million liters (1.1 million tons) from 2018 due to newly implemented provincial and municipal government directives for expanding fuel ethanol blending into gasoline supplies. All fuel ethanol in China's transportation fuel supply must meet national standard GB18350 for denatured fuel ethanol. However, China does not administer a national standard for ethanol inclusion rates in retail gasoline formulations. Industry sources report that although many provinces and cities have adopted E10 in full, actual compliance and the standardization of retail gasoline fuel formulations vary greatly within a given transportation fuel market. As a result of these differences, China's 2019 national fuel ethanol blend rate is forecast at 2.5%, up from 2018, yet slightly lower than peak fuel blend rate levels of 2.5 to 2.8% achieved 10 years ago. MTBE competes with ethanol as a gasoline oxygenate for improving engine performance in China. Recent regulations under the "Blue Sky Protection Plan (2018-2020)" limit the use of MTBE, which further supports expanded ethanol demand. China is not producing ethanol-derived bio-ETBE (ethyl tert-butyl ether) in large scale. Fuel blending formulations incorporating ETBE need additional processing, which have not been used in China. 2019 fuel ethanol production is forecast to jump to a record 4,311 million liters (3.4 million tons), up 1,412 million liters (1.1 million tons), due to new production facilities beginning operations in 2018 and 2019. Operating capacity remains below name plate capacity as existing facilities struggle to manage high production costs, and new facilities progress towards full operating capacity. Industry sources report that an additional 2 facilities or about 258 million liters (204,000 tons) of capacity will launch in 2019, raising the number of China's licensed bio-based fuel ethanol processors to 14, and total production capacity to 5,258 million liters (4.1 million tons) (apps.fas.usda.gov›report›downloadreportbyfilename).

10.4 United States

The United States has the world's fastest growing and largest fuel ethanol market. Figure 10.1 shows Fuel Ethanol Biorefineries by State and Historic Fuel Ethanol Production in United States. Inspite of strong headwinds, the United States ethanol industry powered forward in 2018, breaking records in production, total consumption, and exports. The 210 ethanol plants located across 27 states in United States produced an amazing 16.1 billion gallons of fuel ethanol. Total consumption increased to 16.2 billion gallons, 300 million gallons more than a year ago, driven mostly by record exports of over 1.6 billion gallons as global octane demand continues to grow. Domestic demand also appeared destined for record levels as January blend rates topped a record 10.75%, shattering the so-called blend wall. But as word of EPA's indiscriminate use of small refinery exemptions from the federal Renewable Fuel Standard (RFS) spread, domestic use dropped sharply, hitting a four-year low of just 9.54% in April 2018. Unsurprisingly, such a rapid decrease in demand caused an equally harsh impact on ethanol prices and the profitability. Ethanol prices dropped to a 13-year low as the market responded to preserve some appearance of market share. Reduced demand also had an effect on farmers, who lost a large portion of their single most important market for corn. Net farm income dropped to its lowest level in a decade. Refiners got benefitted. Average monthly United States refiner cash

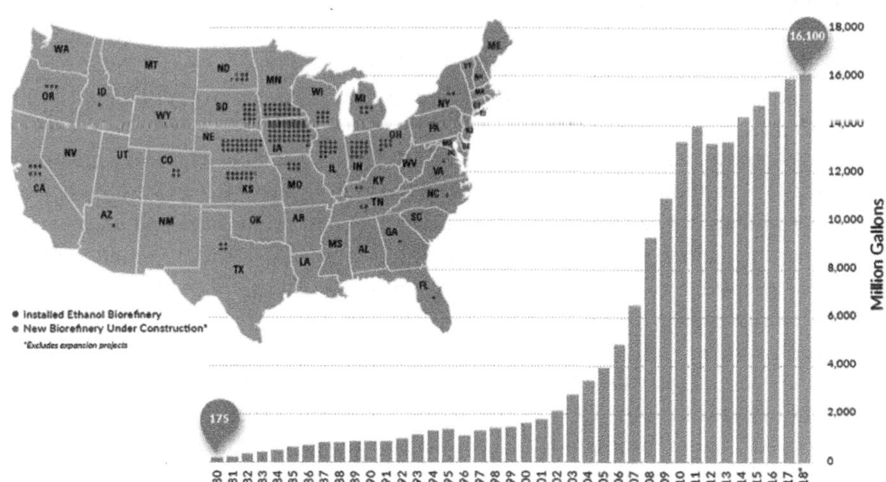

Fig. 10.1 U.S. Fuel Ethanol Biorefineries by State and Historic U.S. Fuel Ethanol Production

operating margins increase gradually throughout the year as ethanol stocks increased and RFS credit prices dipped. Still, things appear more optimistic. President Trump has assured that EPA will communicate a rule allowing the year round use of E15, trade markets are expected to grow even more as uncertainty over several trade agreements dispels and octane demand worldwide continues to rise, and a renewed interest in low carbon fuels predicts bigger requirement for renewable fuels here and abroad. Production from new plants and completed expansions will be available for meeting new requirement and provide a ray of hope for a brighter future. "The use of ethanol in gasoline in 2018 reduced CO_2-equivalent greenhouse gas emissions from the transportation sector by 55.1 million metric tons. That's equivalent to removing 11.7 million cars from the road for an entire year, or eliminating the annual emissions from 13 coal-fired power plants "(2019 ethanol industry outlook—Renewable Fuels Association ethanolrfa.org›wp-content›uploads›2019/02).

Fuel ethanol is consumed across the country and blended in 30% of the gasoline consumed in USA. Earlier, ethanol was used in niche markets in the Midwest, where the production is still concentrated. It is expected that fuel ethanol would be blended in 40% of the gasoline consumed. Ethanol is sold in most states as octane enhancer or oxygenate blended with petrol, and currently covers close to 3% of the USA's gasoline demand. In recent years fuel ethanol demand has been stimulated by the phasing-out of MTBE as octane enhancer. A major concern with MTBE is water contamination and its health effects.

In 2006, the cost of production of fuel ethanol from corn in United States was estimated in the 0.33 to 0.50 Euro/l range against 0.21 to 0.29 Euro/l for cost of production from sugarcane in Brazil (Worldwatch Institute 2006). The energy balance of ethanol production from corn is also much less favourable than in Brazil. The United States Department of Energy and the National Corn Growers Association are cooperating to promote the development of refuelling infrastructure for E85 and to encourage fleet operators to choose ethanol to meet the alternatively fuelled vehicles requirements of the Energy Policy Act. As previously stated, E85 blends require FFT. The potential phase-out of MTBE and an increasing emphasis on domestic energy supply and energy security are likely to favour increased use of fuel ethanol in the United States.

"Consistent with record ethanol production, the United States industry generated a record 41.3 million metric tons (mmt) of distillers grains (DDGS), gluten feed, and gluten meal. These ingredients are mixed into rations for beef and dairy cattle, swine, poultry, and even fish. The industry also produced 4 billion pounds of corn distillers oil, used as a feed ingredient or biodiesel feedstock" https://ethanolrfa.org/wp-content/uploads/2019/02/RFA2019Outlook.pdf.

Ethanol's Value-Added Proposition is shown in the Table 10.4. Based on average prices and product yields in 2018, a typical dry mill ethanol plant was adding nearly $2 of additional value—or 55%—to every bushel of corn processed (ethanolrfa.org).

Figure 10.2 shows production of US ethanol feed co-products (RFA 2018).

Despite the strong interests of corn producers, the long-term sustainability of fuel ethanol production in United States will ultimately depend on the use of new

Table 10.4 Value of outputs per bushel

Ethanol	$3.84
Distillers Grains	$1.16
Corn Distillers Oil	$0.19
Total	$5.19

Corn cost per bushel $3.35

feedstocks and, in this sense, there is a strong commitment to develop new routes of liquid fuels production from cellulosic material.

Ethanol production has plummeted to a six-and-a-half year low in the United States.

According to data from the Energy Information Administration that has been seen by the Renewable Fuels Association, production has fallen by 16.4%, or 165,000 barrels per day to a total of 840,000 barrels per day. This is the largest drop since the Administration began reporting on ethanol production statistics 10 years ago. Ethanol stocks rose 6.5% to a record 25.7 m barrels, eclipsing the previous high set just four weeks ago.

The volume of gasoline supplied to the United States market, which is a measure of implied demand, fell to 6.6 m barrels per day, which was 24.6% lower than the previous week and 27% lower than the same week last year. This fall is reflected in the impact on the industry by the Covid-19 global pandemic and the effect social distancing and stay-at-home policies were having on the market (https://biofuels-news.com/news/us-ethanol-production-crashes-to-six-year-low-new-figures-show/).

Figure 10.2 shows Historic United States ethanol import and exports (RFA 2013) and Fig. 10.3 shows Ethanol production by type (RFA 2010)

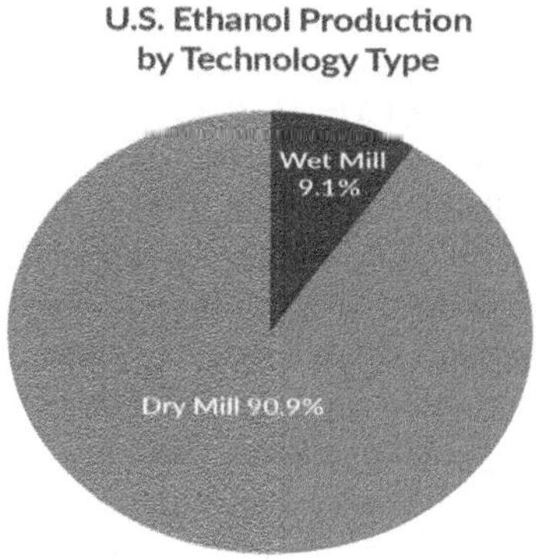

Fig. 10.2 Ethanol production by Technology type, RFA 2020, Ethanol Industry Outlook, https://ethanolrfa.org/wp-content/uploads/2020/02/2020-Outlook-Final-for-Website.pdf Reproduced with permission

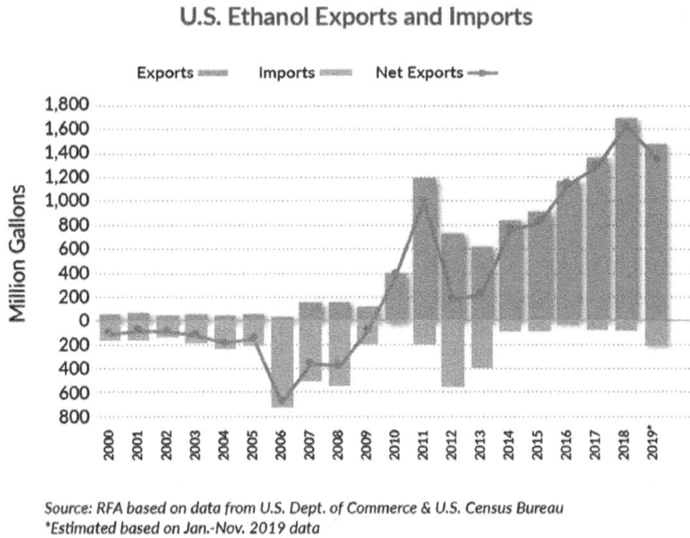

Fig. 10.3 Historic US ethanol import and exports. Reproduced with permission https://ethanolrfa.org/wp-content/uploads/2020/02/2020-Outlook-Final-for-Website.pdf

10.5 Brazil

Brazil has been for decades the world's largest producer and consumer of fuel ethanol, but was overtaken by United States in 2006. Brazil is the world's top exporter; Brazil was the first country to produce bioethanol on a commercial scale, via their government's *Proalcool* programme that was started in 1975 for using fuel alcohol from sugar cane as a gasoline substitute in response to increasing prices of oil. Brazil is now the world's second largest producer in the world and is the largest exporter of fuel ethanol.

Brazil's total ethanol production for 2019 in Brazil was estimated at 34.45 billion liters, an increase of four percent in comparison to the revised figure for 2018. Total ethanol production from corn in 2019 was estimated at 1.4 billion liters, an increase of 609 million liters in comparison to 2018. Total cellulosic ethanol production was estimated at 45 million liters, and represents an unimportant share of total ethanol production in Brazil. No noteworthy changes have been made to the current status of advanced biofuels research and development and production. Total 2019 domestic demand for ethanol (fuel and other uses) was estimated at 33.93 billion liters, up 2.19 billion liters compared to revised figure for 2018. Ethanol is mostly used for fuel in Brazil. Brazil's total ethanol exports were estimated at 1.8 billion liters, an increase of eleven percent in comparison to total exports in 2018 (1.62 billion liters). Brazil's total ethanol imports for 2019 were 1.2 billion liters, a reduction of 495 million liters relative to the earlier year (1.695 billion liters). Ethanol imports are only for fuel use and originate almost totally from the United States.

No changes have been made to the current ethanol mandate, which remains at 27% (E27) for Gasoline C since March 16, 2015. No changes have been made on CIDE and PIS/COFINS for ethanol or gasoline. "The Brazilian government instituted an annual TRQ of 600 million liters on ethanol imports from September 1, 2017 through August 31, 2019. Any volume above the quota is subject to the 20% Common External Tariff under the Mercosur agreement. The Brazillian government has not released any official communication regarding the removal of the TRQ and/or the import tariff on ethanol" (usdabrazil.org.br).

For 2019, total ethanol production was estimated at 34.445 billion liters. This shows an increase of four % in comparison to the revised figure for 2018 (33.149 billion liters). Total ethanol production for fuel use is estimated at 31.387 billion liters, up three % from the preceding calendar year. Total ethanol production from corn in 2019 was estimated at 1.4 billion liters, an increase of 609 million liters in comparison to 2018. Ethanol from corn represents four % of total ethanol production. There are presently ten plants which produce ethanol from corn in Brazil in the states of Mato Grosso and Goias and four plants under construction which should start operations in 1–2 years. The majority of the units are full-plant type, example dedicated corn-only. Some are flex-plants, producing ethanol from sugarcane and corn. There are two new projects recently launched and at least five projects either in the licensing or financing stage. According to the Corn Ethanol National Union (UNEM) Brazil is producing 2.6 billion liters of corn ethanol in 2020 and is expected to produce 8 billion by 2028. Total cellulosic ethanol production for 2019 was at 45 million liters, an increase of 20 million liters in comparison to 2018, assuming that the existing plants are able to overcome the ongoing plant-level operational/mechanical challenges. This amount still represents an unimportant share of total ethanol production in Brazil.

"The total number of sugar-ethanol mills in 2019 is 370 units, one additional unit in comparison to the revised figure for 2018 (369 units). Hydrated ethanol production capacity for 2019 is reported unchanged from the revised figure for 2018 at 43.105 billion liters. This figure reflects the authorized hydrated ethanol production capacity of 233,000 m3 per day, as reported by ANP, and assumes an average of 185 crushing days. Ethanol installed industrial capacity depends on annual decisions made by individual plants to produce sugar and/or ethanol. Post contacts report that the industry responds to the ratio of 40:60 to switch between sugar and ethanol production or vice versa from harvest to harvest. Once producing units adjust their plants to produce a set ratio of sugar/ethanol in a given year, there is much less flexibility to change it during the crushing season"(Brazil Biofuels Annual 2019—USDA Foreign Agricultural Service apps.fas.usda.gov›report›downloadreportbyfilename).

The number of plants producing bioethanol from sugarcane in Brazil is expected to increase to more than 400 in the next few years (Amorim Basso and Lopes 2009; Basso and Rosa 2010). In Brazil, ethanol blends are compulsary (E20 to E25) and anhydrous ethanol is also available from thousands of filling stations. Besides, there are 6 million flex-fuel vehicles in Brazil and 3 million are able to run on E100. In Brazil, Bioethanol now accounts for about 50% of the transport fuel market, where gasoline may now be regarded as the "alternative" fuel.

In Brazil government is encouraging the ethanol industry. A mandate requires all Brazilian gasoline to contain 25% ethanol and the Government helped to fund the establishment of the ethanol supply infrastructure. Brazil's current very strong position regarding fuel ethanol can mainly be explained by the early introduction of a large ethanol support programme in 1975, favourable feedstock production conditions and the wide-spread use of flex-fuel cars. Current support instruments include blending provisions, minor mineral oil tax reductions for fuels containing ethanol and motor vehicle tax reductions for ethanol-powered cars. Fuel ethanol prices have been very competitive compared to petrol prices. Brazil has the world's least cost of ethanol production (Salomao 2005). Energy balances for the production of bioethanol in Brazil are quite impressive.

The blending of bioethanol into gasoline started with the Proálcool program after the first oil crisis in 1975. Today, there is no ethanol-free gasoline on the Brazilian fuel market. All gasoline is marketed with a 25% share of bioethanol (E25, also called gasohol. Actual ethanol content varies as it is yearly adopted to the market situation) and pure bioethanol is on the market as well.

Brazil is the only country in the world with the conditions to quite considerably expand its capacity of ethanol production in short-to mid-term.

10.6 Canada

Canada currently produces relatively low volumes of fuel ethanol. A number of initiatives are underway to boost production significantly. Production is expected to increase significantly in the next few years if current and announced biofuels programmes are implemented. To meet Kyoto Protocol commitments, the country aims to replace 35% of its gasoline use with E10 blends, requiring production of 350 million gallons of ethanol. Seven new plants with total capacity of 200 million gallons are planned under the Ethanol Expansion Program. Ontario, Saskatchewan, and Manitoba are already promoting ethanol through production subsidies, tax breaks, and blending requirements. In Canada, ethanol is produced almost entirely from cereals.

Gasoline mixed with ethanol is now available at over several gas bars across Canada from Quebec to the Pacific, including the Yukon Territory. In several regions, ethanol blends are available for bulk delivery for farm and fleet use. The federal government and many provinces are offering tax incentives.

USDA Foreign Agricultural Service's Global Agricultural Information Network provides an overview of biofuel industry of Canada and the country's changing biofuels policy framework (https://www.fas.usda.gov›data›canada-biofuels-annual-5). Canada is at present in the process of developing a Clean Fuel Standard, which will enact a carbon intensity strategy when accounting for the amount of blended renewable fuel. Once finalized, the Clean Fuel Standard will replace the volumetric approach presently in place under federal renewable fuels regulations of Canada.

Environment and Climate Change Canada released the regulatory design paper for the Clean Fuel Standard in December 2018. There are separate requirements for liquid, gaseous and solid fuels. The ECCC released a benefit-cost analysis framework for the Clean Fuel Standard in February 2019. Information revealed on the Clean Fuel Standard regulations to date shows the program will reduce the carbon intensity of liquid fuels to be reduced by 10 grams of carbon dioxide equivalent per megajoule below the reference carbon intensity (year 2016) by 2030.

Under the proposed CFS, separate carbon intensity requirements would be established for subsets of fuels in the following sectors: transportation, building requirements, and industry," said the authors in the report. "The proposed CFS will not differentiate between crude oil types that are produced domestically or are imported. The federal government will maintain national blending mandates in the short-term, establishing an 'expiration date' for the volumetric requirements through consultations with stakeholders.

Proposed Clean Fuel Standard regulations for the liquid fuels are published in 2019. Regulations for gaseous and solid fuel streams will be established at a later date.

At present, federal renewable fuels regulations in Canada need fuel producers and importers to have an average ethanol content of at least 5% based on the volume of gasoline produced or imported. The federal regulations also need 2% renewable content based on the volume of diesel fuel and heating distillate oil fuel producers and importers produce or import.

Several provincial blend mandates are also in place (Table 10.5). Manitoba has an 8.5% ethanol blend mandate for gasoline, whereas Saskatchewan has a 7.5% mandate and British Columbia, Alberta, and Ontario each have a 5% mandate. For renewable blends in diesel, British Columbia and Ontario have a 4% mandate, whereas Manitoba, Alberta and Saskatchewan each have a 2% mandate.

Canada produced 1.83 billion liters of fuel ethanol this year, up from 1.75 billion liters in 2018. Imports are expected to reach 1.37 billion liters this year, down from 1.39 billion liters in 2018. Fuel ethanol consumption is expected to reach 3.2 billion liters this year, up from 3.14 billion liters in 2018. The United States supplies essentially all of Canada's fuel ethanol imports.

Table 10.5 Provincial Blend Mandates

Province	Ethanol Blend Mandate for Gasoline
British Columbia	5%
Alberta	5%
Saskatchewan	7.5%
Manitoba	8.5%
Ontario	5%
Quebec	No blend requirement

https://ethanolrfa.org/wp-content/uploads/2019/02/RFA2019Outlook.pdf

Canada presently has 13 ethanol plants in operation, a number that has been stable since 2015. Nameplate capacity, however, is up. Capacity is expected to reach 2.15 billion liters this year, up from 1.97 billion liters in 2018 and 1.872 billion liters in 2017. Capacity use is expected to be at 85% this year, down from 89% in 2018 and 92% in 2017.

The overall ethanol blend rate in Canada is expected to reach 6.6% this year, up from 6.4% in 2018 and 6.2% in 2017.

"Greenfield Global, the largest producer of ethanol in Canada, has taken steps for expanding production at its Varennes biorefinery following the introduction of draft regulations on the minimum renewable fuel volumes in the province of Quebec. Welcoming the Government of Quebec's draft regulations on the minimum volume of renewable fuel to be blended in gasoline and diesel, the producer added that it could result in a major expansion of its production in the province. The new proposal would set blending thresholds of 10% renewable fuel in gasoline, and 2% renewable fuel in diesel by 2021. This would increase to 15% in gasoline and 4% in diesel by 2025" (gain.fas.usda.gov).

"Greenfield applauds the Quebec Government for its ambitious targets and firm commitment to transition the province's energy," said Jean Roberge, executive vice-president and managing director of renewable energy at Greenfield. "Biofuels are indeed a sector for the future. The government is demonstrating true leadership on climate change and taking steps necessary to effectively and efficiently increase the use and production of renewable fuels, like ethanol, in the province."

Greenfield's Varennes biorefinery, which is the first ethanol plant in Quebec, has been in operation since 2007. The facility produces the lowest carbon intensity ethanol in Canada, and also corn oil and distillers' grain.

The company is presently considering adapting emerging advanced biofuels technologies using non conventional feedstocks and processes, including cellulosic ethanol, renewable diesel and renewable natural gas, to further reduce the carbon intensity of its biofuels.

The first phase of Greenfield's expansion study was completed last year, with the second now due to commence with engineering, feedstock and environmental studies.

10.7 India

India is one of the world's largest sugar producers. In February 2009, India and the United States exchanged a memorandum for cooperation on biofuels development, covering the production, utilization, distribution and marketing of biofuels in India. Traditionally most of the Indian ethanol production is directed to the industrial consumption. Recently, mainly due to economic and strategic reasons, Indian government has seriously considered fuel ethanol production and a mandate for E10 blends is currently effective in 13 states. In the near future, the mandate for E10 shall be effective in the whole country. In addition, the Indian Institute of Petroleum had conducted experiments using a 10-percent ethanol blend in gasoline and 15-percent ethanol in biodiesel.

USDA Foreign Agricultural Service's Global Agricultural Information Network report shows the average ethanol blend rate in India will reach 5.8% this year, higher from 4.1% set last year (http://ethanolproducer.com/articles/16451/report-india-to-use-produce-a-record-volume-of-ethanol-in-2019).

India is presently aiming to obtain an E10 blend by 2020 and E20 by 2030. Ethanol Blending Program specifies procuring ethanol produced from molasses, sugarcane juice, and damaged food grains. A surplus sugar season coupled with financial incentives for converting excess sugar into ethanol would help the oil marketing companies procuring over 2.4 billion liters of ethanol this year.

It is doubtful that the India's E20 objectives will be achieved by 2030 because of the general incapability of the cane industry; it is not able to supply fuel demand of India, the fact that imports are managed in a way which reduces the role they can play, and the expected period for large scale production of advanced biofuels.

India will consume 3.8 billion liters of ethanol this year, up from 3.1 billion liters in 2018. A 6.6% blend rate could be achieved if all the ethanol produced from molasses this year is blended with gasoline. "Potential blending would be higher yet if imports were permitted and duties lowered," the report states. "However, given demand from the potable and industrial sectors and limitations on imports, a national blend average of 5.8% in 2019 is expected."

Ethanol production in India is expected to reach 3 billion liters this year, higher by 11% from 2018. Last year, about 2.7 billion liters of ethanol was produced from molasses. As regards imports, the United States has remained the largest ethanol supplier to India for the last six years. Indian ethanol importers were down 14% last year, declining to 633 million liters. The United States accounted for 94% of 2018 imports (http://ethanolproducer.com/articles/16451/report-india-to-use-produce-a-record-volume-of-ethanol-in-2019).

India had 166 ethanol refineries last year, up from 161 in 2018. That number is expected to increase this year. Capacity use was 117% in 2018 and is expected to fall go 115% in 2019.

India is strengthening its bioethanol program by analyzing and mixing of a variety of resources, price, and raw material. The Ministry of Petroleum and Natural Gas of the Government of India has authorized 5% ethanol blend in gasoline (Jahnavi et al. 2017).

India is moving steadily toward the dependency and use of biofuels (Bandyopadhyay 2015). The Centre for Biofuels, under CSIR, is developing the key phase of a research project for developing technology for producing bioethanol as an alternative fuel for transportation. Among the biomass residues which can be possibly used as substrate, 80% can be produced mostly from crops (survey performed by IMRB, employed by National Institute for Interdisciplinary Science and Technology (NIIST; Jahnavi et al. 2017). A pilot plant producing ethanol from biomass, has been set up on the NIIST campus. A multi feedstock plant will be designed to use different types of raw materials for producing ethanol. Based on a study covering several thousand farmers, the NIIST has highlighted the surplus biomass resources available in different parts of the country for producing bioethanol. "Swachh Bharat Abhiyan, Department of Biotechnology, Ministry of Science and Technology-supported DBT-Institute of Chemical Technology (ICT) Center for Energy (Mumbai), has successfully demonstrated commercial lignocellulosic bioethanol production for the first time. Glycols Ltd. (Kashipur, Uttarakhand) was established as India's first cellulosic ethanol production plant with a production capacity of 10 t/day. The plant was successfully operated with various lignocellulosic feedstock such as wheat straw, rice straw, bagasse, cotton stalk, and bamboo and has a potential to convert those lignocellulosic residues into ethanol in less than 24 h. The DBT-ICT Center for Energy is confident about their newly developed plant design that can convert 250 and 500 t/day biomass to 2G-Alcohol and can be sold at a competitive price. Although there has been a massive growth in lignocellulosic bioethanol production in India, still membrane-integrated processes have been rarely investigated, more specifically in the industrial level for lignocellulosic bioethanol production" (www.degruyter.com).

10.8 Thailand

The Thai government is pursuing a policy of increasing consumption of biodiesel and ethanol produced domestically from cassava. One of its initial steps is to replace the octane-enhancing additive MTBE in gasoline with ethanol. Large-scale production of fuel ethanol production has started with molasses but cassava was officially designated the prime raw material. Thailand is a large producer of cassava. As the domestic price of sugar is too high it doesn't seem worth to produce ethanol from sugarcane. In Thailand all premium gasoline shall be replaced by E10.

According to report from the United States Department of Agriculture's Foreign Agricultural service, Thailand is relying upon domestic ethanol production, the country could get advantage from importing biofuels.

Whereas the country allows imports of ethanol for industrial application, the country is not importing ethanol for use as a transportation fuel. Traders of ethanol in Thailand are needed to receive a permit from Ministry of Energy; but, "to date, the MOE has never approved any imports of fuel ethanol due to sufficient supplies of locally produced ethanol".

Raw materials for domestic ethanol production in Thailand are sugar cane, molasses and cassava. Due to shortages of these feedstocks, Thailand "will be forced to temporarily lower biofuel use targets or price surges when weather-related feedstock shortages occur," and the country's lack of ethanol imports will stop it from meeting higher targets of ethanol use. It is also expected to see higher greenhouse gas emissions from land use change. "Permitting some role for imports unlocks the full positive potential contribution biofuels can make."

The country is set to reduce it's consumption goals for 2037. A new 20-year Alternative Energy Development Plan has been approved and "the government is in the process of reviewing ethanol and biodiesel consumption targets." Those targets are expected to be lesser than the targets set out in the 2015 plan, due to the concerns over raw material supply. The new targets are expected to reduce the consumption target in 2037 to 2.4 billion liters which is 41% below the target set out in 2015.

The ethanol and production rates in Thialand are expected to increase in 2019, but not as fast as they once had. In 2017, the country observed its highest ethanol consumption growth rate in 2017 at 12%. In 2019, ethanol consumption has increased by 6% over 2018 levels. The decreased rate of ethanol production and use is "due to the delay in the cessation of Octane 91 E10 sales." Inspite of the delay, the ethanol-blend levels in the country have reached 13.5% in this year as a result of strong E20 sales (http://www.ethanolproducer.com/articles/16741/report-thailand-could-benefit-from-ethanol-imports).

10.9 Japan

Japan is one the main consumers of motor gasoline in the world and is heavily dependent on imported oil. The country has considered large-scale use of fuel ethanol, or ETBE, targeting at improving its energy security and at reducing GHG emissions, in this case in order to accomplish its Kyoto obligations. By now ethanol blends are used in some regions. Japanese government intends to define a mandate for E3 valid for the whole country and to expand this mandate to E10 in next few years. However, there is some resistance due to the low number of large-scale ethanol suppliers and also due to the interests of oil companies that prefer gasoline blends with ETBE rather than with fuel ethanol (Piacente 2006).

In Japan, consumption of gasoline in 2018 was 51 billion liters. "The pure ethanol equivalent of ETBE consumed, plus a small amount of ethanol consumed in direct blending, brings total fuel ethanol consumption to 890 million liters, and yields an effective national average blend rate of 1.8%" (gain.fas.usda.gov).

Japan has one refinery which is producing about 0.2 million liters of bioethanol for fuel application from domestic rice. No change is anticipated to this level of production now. The refinery is located in Niigata Prefecture and is operated by JA Zen-noh, the federation of agricultural cooperatives. It uses high yield rice grown especially for biofuel production. The ethanol is used as part of an E3 blend, and the E3 gasoline is sold at six affiliated gas stations in the region of Niigata Prefecture.

In 2010, Japan Biofuels Supply LLP started producing ETBE domestically. Each year, the company is producing 170 million liters of ETBE, using 72 million liters of ethanol. The company is fully dependent on imported ethanol. Oil industry of Japan is now allowed to use United States ethanol for producing ETBE for meeting Japan's biofuels mandate (gain.fas.usda.gov›BiofuelsAnnual_Tokyo_Japan_11).

References

Amorim HV, Basso LC, Lopes ML (2009) Sugar cane juice and molasses, beet molasses andsweet sorghum: composition and usage. In: The alcohol textbook, 5th edn. Nottingham University Press, pp 39–46

Bajpai P (2007) Bioethanol. PIRA Technology Report, Smithers PIRA, UK

Bandyopadhyay K (2015) Biofuel promotion in India for transport: exploring the grey areas. Policy Br 2015: 1–12

Basso LC, Rosa CA (2010) Sugar cane for potable and fuel ethanol. In: Proceedings of the worldwide distilled spirits conference, Edinburgh, 2008. Nottingham University Press

Goldemberg J (2008) The Brazilian biofuels industry. Biotechnol Biofuels 1:6–6

Jahnavi G, Prashanthi GS, Sravanthi K, Rao LV (2017) Status of availability of lignocellulosic feed stocks in India: biotechnological strategies involved in the production of bioethanol. Renew Sustain Energy Rev 73:798–820

Johnson T, Johnson B, Scott-Kerr C, Kiviaho J (2010). Bioethanol—status report on bioethanol production from wood and other lignocellulosic feedstocks. In: 63rd appita annual conference and exhibition, Melbourne, 19–22 April 2009

Piacente EA (2006) Perspectives for Brazil in the bio-ethanol market. MSc dissertation. State University of Campinas—Unicamp, Campinas

RFA-Renewable Fuels Association (2018) Ethanol strong, Ethanol industry outlook. https://ethanolrfa.org/wp-content/uploads/2018/02/NECfinalOutlook.pdf

Salomao A (2005) O novo siclo da cana de acucar. Exame, ed. 845 No. 12, 22 June

Yan X, Inderwildi OR, King DA (2010) Biofuels and synthetic fuels in the US and China: a review of well-to-well energy use and greenhouse gas emissions with the impact of land-use change. Energy Environ Sci 3:190–197

Zhenhong Y (2006) Bio-fuels industry in China: utilization of ethanol and biodiesel in today and future. In: World biofuels symposium, Beijing, China

Relevant Websites

http://ethanolproducer.com/articles/16451/report-india-to-use-produce-a-record-volume-of-ethanol-in-2019

http://www.ethanolproducer.com/articles/16741/report-thailand-could-benefit-from-ethanol-importsethanolproducer.com›articles›iea-predicts-growth-in.gain.fas.usda.gov›BiofuelsAnnual_Tokyo_Japan_11-

https://biofuels-news.com/news/us-ethanol-production-crashes-to-six-year-low-new-figures-show/

https://ethanolrfa.org/wp-content/uploads/2019/02/RFA2019Outlook.pdf

https://ethanolrfa.org/2019/02/rfa-releases-2019-ethanol-industry-outlook-pocket-guide/

https://ethanolrfa.org/wp-content/uploads/2019/02/RFA2019PocketGuide.pdf.apps.fas.usda.gov›report›downloadreportbyfilename

https://apps.fas.usda.gov/newgainapi/api/report/downloadreportbyfilename?filename=Biofuels%20Annual_Beijing_China%20-%20Peoples%20Republic%20of_8-9-2019.pdf

https://noaw2020.eu/wp-content/uploads/2018/12/C1-1-COFCO-Wuguoqing.pdf

https://arena.gov.au/assets/2019/11/biofuels-and-transport-an-australian-opportunity.pdf

https://www.e-education.psu.edu/egee439/node/646

https://www.statista.com/statistics/281606/ethanol-production-in-selected-countries/

https://ethanolrfa.org/statistics/annual-ethanol-production/

Status and progress of fuel ethanol in China. COFCO. Beijing https://noaw2020.eu/wp-content/uploads/2018/12/C1-1-COFCO-Wuguoqing.pdf

https://www.fas.usda.gov/data/eu-28-biofuels-annual-1

http://usdabrazil.org.br

Chapter 11
Ethanol and Environment

Abstract Environmental impact of bioethanol production technologies and their life-cycle assessment are discussed. Energy production and utilization cycles based on cellulosic biomass have near-zero green house gas emissions on a life cycle basis. Use of biomass for ethanol production offer environmental advantages in terms of nonrenewable energy consumption and global warming impact.

Keywords Biofuels · Bioethanol · Fuel ethanol · Life-cycle assessment · Environmental benefits

Ethanol is a significantly important chemical. In actual fact, ethanol has been used for eight thousand years or so. Some research show that Stone Age people identified the importance of a good-quality alcoholic drink (Brick 2005).

Some still argue that "ethanol, in the form of beer and wine, was an evolutionary driver in helping humanity transcend its original hunter-gatherer lifestyle and begin living in denser population clusters" (Will 2008). The potential of ethanol as motor vehicle fuel has long been recognized by automobile designers. The first American internal combustion prototype was produced by Samuel Morey in 1826; it was able to run on ethanol, and remained the leading automotive fuel until 1908, when a combination of quickly decreasing price of gasoline and swiftly increasing price of ethanol led the Ford Motor Company to introduce the Model T, which is an early "flex-fuel" vehicle which could use either ethanol or gasoline. In the United States blends of gasoline and ethanol were used until the 1920s and in Europe until the 1930s but were ultimately replaced by gasoline containing lead additives, which was found to prevent engine knocking. Ethanol blends were in actual fact off the market by 1940 (Cole 2006).

Ethanol weakened in niche applications for the next three decades. In 1973, the oil embargo made ethanol appear attractive again as a substitute for gasoline and renewed an awareness in building ethanol distilleries. Gas lines and shortages in the 1970s even prompted people to think about fermenting waste of fruits and vegetables and surplus to widen their supplies of gasoline (Langley 1979). When the gasoline price dropped in the 1980s, interest in ethanol reduced, but ethanol was given another look in 1989 when air pollution regulators approved the use of fuel oxygenates (including

P. Bajpai, *Developments in Bioethanol*, Green Energy and Technology, https://doi.org/10.1007/978-981-15-8779-5_11

ethanol) for reducing pollution (Green 2006). This amused the farm lobby, which was taking taxpayer subsidies for turning a crop surplus into fuel supplements required by air quality laws. "Later discovery of the groundwater-polluting effects of methyl tertiary butyl ether (MTBE)—the oxygenate that was ethanol's major competitor—led several states to switch from MTBE to ethanol as a fuel oxygenate. Recent developments have sent the fortunes of ethanol producers skyrocketing, as ethanol fuel has come back with a subsidy-powered roar" (www.eia.doe.gov/emeu/steo/pub/special/mtbe).

"While ethanol promoters make it sound as if ethanol is the solution to all our energy woes—dependence on foreign oil, diminishing oil stocks, the environmental consequences of energy use, the decline of the family farm, and so on—a considerable amount of research has shown that ethanol has far more peril than it does promise" (witsendnj.blogspot.com).

While ethanol is generally considered as a proper solution to climate change as it merely recirculates carbon in the atmosphere, there is more than one type of greenhouse gas (GHG) to consider. Ethanol, mixed with gasoline, in point of fact increases the release of strong greenhouse gases more than gasoline. Since 1997, the United States Government found that the ethanol production process generates comparatively more nitrous oxide and other strong greenhouse gases as compared to gasoline. On the contrary, the GHG emitted during the traditional gasoline fuel cycle contain comparatively more of the less strong type, carbon dioxide (Bryce 1997).

Although the United States Environmental Protection Agency (US EPA) claims a net reduction in GHG emissions from using ethanol, they found that use of ethanol is a issue for conventional air pollutants. Ethanol use, will increase the release of chemicals which produce ozone which is one of the most tough air pollutant. Concomitantly, other vehicular emissions may increase due to higher use of renewable fuel. All over the country, an increase in total emissions of volatile organic compounds (VOCs) and nitrogen oxides (NO) between 41,000 and 83,000 tons because of higher use of ethanol is estimated by EPA. Areas which see a significant increase in ethanol may notice an increase in volatile organic compounds emissions between 4 and 5% and an increase in NO emissions between 6 and 7% from gasoline powered (United States Environmental Protection Agency 2007).

According to estimate by Mark Z. Jacobson, (Stanford University), that "switching to a mixture of 85% ethanol and 15% gasoline—compared to 100% gasoline—may increase ozone-related mortality, hospitalization, and asthma by about 9% in Los Angeles and 4% in the United States all together" (Jacobson 2007).

Almost all forms of energy production use a substantial amount of water (Kreider and Curtiss 2007). "The amount of water consumed in ethanol production has dropped significantly as ethanol production has become more efficient. Today, it takes about 3.5 gallons of water to produce one gallon of ethanol—and much of that is processed and returned to streams and watersheds. At one time, it took 8 gallons of water to produce one gallon of ethanol, so the industry is continually getting better at conserving this precious resource. To put this in perspective, a 100 million gallon per year ethanol plant uses about as much water as it takes to irrigate about 1,000 residential lawns of 5,000 square feet each during a typical spring-summer season" (CFDC 2010).

According to National Academy of Sciences (2007) that "if the United States continues to increase corn-based ethanol production without new environmental protection policies, the increase in harm to water quality could be considerable. Corn, according to the NAS, requires more fertilizers and pesticides than other food or biofuel crops. Pesticide contamination is highest in the corn belt, and nitrogen fertilizer runoff from corn already has the highest agricultural impact on the Mississippi River. In short, more corn raised for ethanol means more fertilizers, pesticides, and herbicides in waterways; more low-oxygen dead zones from fertilizer runoff; and more local shortages in water for drinking and irrigation".

Ethanol represents closed carbon dioxide cycle as after burning of ethanol, the carbon dioxide which is emitted is recycled back into plant material as plants use carbon dioxide for synthesizing cellulose during photosynthetic cycle (Chandel et al. 2007). Ethanol production process simply uses energy from inexhaustible energy sources; no net carbon dioxide is added to the atmosphere. This makes ethanol an environmentally useful energy source. Additionally, the toxicity of the emissions released from ethanol is lesser in comparison to petroleum sources (Wyman and Hinman 1990). Ethanol produced from biomass does not contribute to the GHG effect (Foody 1988). Because of the increase in energy demand, the worldwide supply of fossil fuels cause damage to human health and also contributes to the GHG emission (Demirbas 2007). Hahn-Hagerdal et al. (2006) "alarmed to the society by seeing the security of oil supply and the negative effect of the fossil fuel on the environment, especially on green house gas emissions. The reduction of green house gas pollution is the main benefit of using biomass conversion into ethanol. Ethanol contains 35% oxygen which helps in complete combustion of fuel and thereby reduces particulate emission which cause health problem to living beings".

Figure 11.1 shows greenhouse gas reductions compared to gasoline. Figure 11.2 shows that ethanol has reduced GHG emissions by 22 MMT in California since 2011.

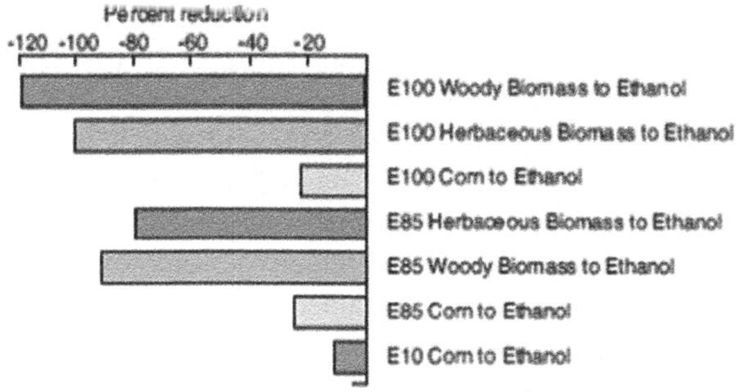

Fig. 11.1 Reductions in greenhouse gas in comparison to Gasoline. Ethanol Fact Book (2016). A compilation of information about Fuel Ethanol. https://4-h.org/wp-content/uploads/2016/02/Ethanol-Fact-Book-1107.pdf. US department of energy office of fuels development and congressional research service

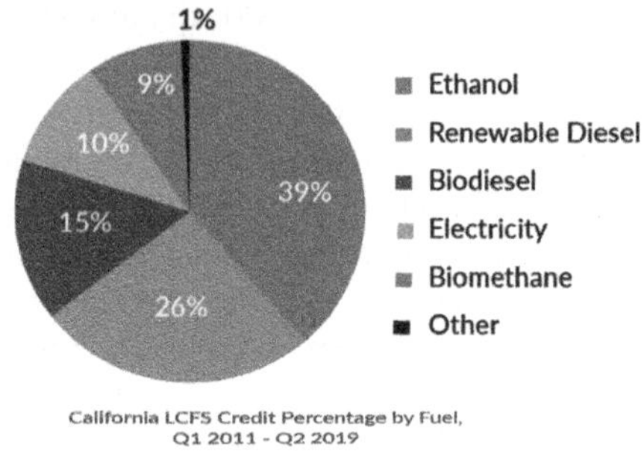

Fig. 11.2 Ethanol has reduced GHG emissions by 22 MMT in California since 2011. Clean Fuels Development Coalition, RFA

Table 11.1 Clearing the air with Ethanol

Carbon monoxide May cause deleterious health effects by reducing oxygen delivery to the body's organs
Exhaust hydrocarbons Contribute to ozone—irritate the eyes, damage the lungs and intensify the respiratory problems
Air toxics like benzene May cause cancer and reproductive effects or birth defects
Fine particulate matter Can pass through the throat and nose and enter the lungs causing serious health effects

https://www.aei.org/research-products/report/ethanol-and-the-environment/

Table 11.1 shows clearing the air with ethanol.

Bang-Quan et al. (2003) performed study on the ethanol mixed diesel (E10 and E 30) combustion at diverse loads. They observed that mixing of ethanol to diesel fuel concurrently reduces cetane number, high heating value, aromatics fractions and kinematic viscosity of ethanol blended diesel fuels and changes distillation temperatures. These factors lead to the complete burning of ethanol and lesser emissions. Bioethanol is able to reduce ozone precursors by 20—30%. Therefore it can play an important role in reducing the damaging gasses in metro cities globally. There is 41% reduction in particulate matter and 5% NOx emission with diesel blended with ethanol (E-15) (Subramanian et al. 2005; Chandel et al. 2007). One of the drawback in using ethanol as fuel is that aldehyde mostly acetaldehydes emissions are higher in comparison to that of gasoline. But acetaldehydes emissions produce less undesirable health effects as compared to formaldehydes released from gasoline engines (Gonsalves 2006).

RFG with Oxygenates, such as ethanol, significantly reduces unsafe gasoline emissions (Table 11.2).

Use of ethanol/gasoline blends up to 10% is approved by automobile manufacturers. Endorsement of ethanol/gasoline blends is found in the owner's manual under references to refueling or gasoline. General Motors Corporation has recommended

Table 11.2 Reduction of harmful gasoline emissions

Air toxics	28%
Volatile organic compounds	17%
Nitrogen oxides	3%
Carbon monoxide	13%
Sulfur oxides	11%
Carbon Dioxide	4%
Particulate matter	9%
Reduced cancer risk	20–30%

the use of fuel oxygenates, like ethanol, when and where available. Fuel ethanol mixtures are sold in almost all the state and can be found in more than 46% of the nation's gasoline. Blend of ethanol and gasoline has obtained almost 100% market share of all gasoline sold in certain carbon monoxide (oxygenated gasoline) and ozone nonattainment areas (reformulated gasoline). Minnesota is using a oxygenated fuel program which has resulted in ethanol being mixed in more than 95% of the state's gasoline. Thus fuel ethanol is effectively used in all types of vehicles and engines which need gasoline.

Ethanol or blends of ethanol have by no means contributed to burning or fouling of port fuel injectors. Some constituents in gasoline, like olefins, cause deposits which may foul injectors. Ethanol does not contribute to the formation of deposits as it burns 100% and no residue is left. Ethanol blends in fact keep fuel injectors cleaner and help in improving the performance of engine. Ethanol does not harms any seals or valves nor increase corrosion.

Vapor pressure specifications of gasoline continue to be reduced by state and federal statute, almost removing the vapor lock problems that were found years ago. In addition, all major auto makers now have in-tank fuel pumps, which are not subject to vapor lock similar to the older in-line fuel pumps. Ethanol actually helps in keeping the engine cooler, as the ethanol in the fuel burns at a lower temperature. Actually, several high powered racing engines use pure alcohol for this reason. The IndyCar Series® converted to using 100% ethanol in 2007 (ethanolacrossamerica.net).

"Life-cycle assessment (LCA) is a conceptual framework and methodology for the assessment of environmental impacts of product systems on a cradle-to-grave basis. Analysis of a system under LCA encompasses the extraction of raw materials and energy resources from the environment, the conversion of these resources into the desired products, the utilization of the product by the consumer, and finally the disposal, reuse, or recycle of the product after its service life. The LCA approach is an effective way to introduce environmental considerations in process and product design or selection" (Chandel et al. 2007; Azapagic and Clift 1999).

The comparison of different ethanol production technologies can be made based on life cycle assessment studies. Energy production and consumption cycles based on cellulosic biomass materials have almost zero green house gas emissions on a

Table 11.3 Estimated greenhouse gas (GHG) emissions (direct and indirect) associated with cellulosic ethanol production from cereal straw and perennial crops

	Direct emissions from RED Recast, Annex V—Default Values (g CO_2eq/MJ)	Indirect emissions (ILUC emissions) (g CO_2eq/MJ)	Indirect emissions (Displacement emissions) (g CO_2eq/MJ)
Ethanol from wheat straw	16	16 if unsustainable straw removal is assumed (more than 33–50%) 0 if sustainable straw removal is assumed (less than 33–50%)	8 displaced uses: Livestock bedding and feed; mushroom cultivation; horticulture; heat and power
Ethanol from perennial grass (miscanthus and swithgrass)	16	– 12	0 no existing uses

Padella et al. (2018). Reproduced with permission

life cycle basis (Table 11.3). Use of biomass for production of ethanol presents environmental advantages in terms of unrenewable energy consumption and global warming impact.

Kim and Dale (2004) studied LCA giving emphasis to use of corn and soyabean production from bioethanol and biodiesel production. They found that both type of biofuels have environmental advantages in terms of unrenewable energy consumption and impacts of global warming. But use of ethanol for biomass production also increases acidification and eutrophication, mainly due to the reason that large amounts of nitrogen, phosphorus are released after the crops are grown.

Lechon et al. (2005) studied the LCA of ethanol production from wheat and barley. Barley is a superior alternative in comparison to wheat in terms of lower emissions of green house gases.

Dissimilar to gasoline, pure ethanol is not toxic and biodegradable, and it rapidly breaks down into non toxic substances if spills out. Chemical denaturants are mixed with ethanol for making fuel ethanol, and several of the denaturants are toxic. Like gasoline, ethanol is a very combustible liquid and should be transported cautiously. https://www.eia.gov/energyexplained/biofuels/ethanol-and-the-environment.php.

Ethanol and mixture of ethanol and gasoline burn cleaner and have higher octane levels as compared to pure gasoline, but they also show higher evaporative emissions from fuel tanks and dispensing equipment. These evaporative emissions contribute to the formation of injurious, ground level ozone and smog. Gasoline needs extra processing for reducing evaporative emissions before mixing with ethanol.

Carbon dioxide, a greenhouse gas is released when ethanol is produced and burnt. But, the combustion of ethanol produced from biomass is believed to be atmospheric carbon neutral because as the biomass grows, it absorbs carbon dioxide, which can counterbalance the carbon dioxide released when the ethanol is burnt. Some ethanol producers burn coal and natural gas for heat sources in the fermentation process for making fuel ethanol, whereas some burn corn stocks or sugar cane stocks.

The Argonne National Laboratory has reported that using "ethanol in gasoline produces 32% fewer emissions of greenhouse gases than gasoline for the same distance traveled. Ethanol also reduces emissions of other harmful pollutants such as carbon monoxide—and it dilutes and displaces toxic gasoline components such as benzene and toluene" (CFDC 2010).

The effect that higher use of ethanol has on net emission of carbon dioxide depends on how ethanol is produced and whether or not indirect impacts on land use are included in the computation. Growing plants for fuel is a debatable topic as few people think that the land, fertilizers, and energy used for growing biofuel crops should be used for growing food crops in its place.

The United States government is supporting efforts for producing ethanol using the methods using lesser energy as compared to traditional fermentation using cellulosic biomass, which needs less cultivation, fertilizer, and pesticides as compared to corn or sugar cane. Raw materials which are used for producing cellulosic ethanol include native prairie grasses, faster growing trees, sawdust, and even waste paper etc.

RFA (https://ethanolrfa.org/environment/) 2014 through 2018 ranked as the earth's five warmest years on record, and 2019 has topped them all. Furthermore, record fires in Australia, record ice melt in the Arctic, and increasingly volatile weather patterns have sparked a worldwide dialogue about the increasing urgency for reducing carbon.

From the viewpoint of RFA, the ethanol industry should help to lead the conversation. The **U.S. Department of Energy**, California Air Resources Board (**CARB**), Oregon Department of Environmental Quality, U.S. Department of Agriculture (**USDA**) and others have already recognized that ethanol produced from grain reduces greenhouse gas (GHG) emissions by 35–50% in comparison to gasoline. GHG emissions would reduce by around 70% in just the next few years, according to USDA. Further, CARB reports that "ethanol is responsible for 22 million metric tons of GHG reduction from California's transportation sector since 2011—more than any other low carbon fuel".

United States already has a framework for driving future policy. The **Renewable Fuel Standard** has been an important and effective policy reducing greenhouse gas emissions from fuels for 15 years. It has "reduced carbon dioxide equivalent GHGs by an astounding 600 million metric tons since its implementation. That is the equivalent of removing roughly half of the cars on the road in America for an entire year or eliminating the annual emissions from 13 coal-fired power plants. With **ethanol**, we do not have to wait and hope for main technological or economic breakthroughs. Ethanol fuel is available now at a low cost to drive decarbonization of our liquid fuels".

Besides reducing emissions of GHG, ethanol is the best option for reducing tailpipe emissions of other harmful pollutants. Addition of ethanol to gasoline reduces tailpipe emissions of the several pollutants.

Table 11.4 Environmental and resource issues of concern with biofuels

RFS2 mandated use of biofuel	
Environmental issues	Resource conservation issues
Air quality, Water quality includes pesticides, sediments, nutrients, pathogens, and acreage/function of wetlands Soil quality	Soil conservation, Water Availability, Ecosystem health and biodiversity includes invasive/noxious plants, forests, grasslands, wetlands, and other aquatic ecosystems

Based on (EPA 2011), (Hoekman et al. 2018)

Table 11.5 Environmental impact of bioethanol technologies

Positives • Uses energy from renewable energy sources No net carbon dioxide is added to the atmosphere; makes ethanol an environmentally useful energy source • Toxicity of exhaust emissions is lesser as compared to that of petroleum sources • Energy crops grown for the producing ethanol absorbs enormous amounts of green house gases emitted by the burning of fossil fuels • 35% oxygen is present in ethanol which helps complete combustion of fuel and thus reduces particulate emission which creates health danger to human beings
Negatives • Deriving ethanol from crops (example—corn) consumes huge amounts of nitrogen fertilizer and extensive top-soil erosion associated with growing of this particular crop

Table 11.4 presents environmental and resource issues of concern with biofuels. In Table 11.5, environmental impact of bioethanol technologies are summarized.

References

Azapagic A, Clift R (1999) Life cycle assessment and multiobjective optimisation. J Clean Prod 7:135–143

Bang-Quan H, Shi-jin S, Jian-Xin W, Hong H (2003) The effect of ethanol blended diesel fuels on emissions from a diesel engines. Paper No. SAE 2003-01-0762, World Congress, Detroit, MI, March 3–6

Brick J (2005). Neuropharmacology of Alcohol, Rutgers University, Center of Alcohol Studies, August 2005. http://alcoholstudies.rutgers.edu/onlinefacts/neuro.html

Bryce R (1997). Gusher of Lies: The dangerous delusions of energy independence. PublicAffairs, New York, NY 2008, 165; and U.S. General Accounting Office, Tax Policy: Effects of the alcohol Fuels Tax Incentives, report to the chairman, House Committee on Ways and Means, March 1997, 17. www.gao.gov/archive/1997/gg97041

Chandel AK, Chan ES, Rudravaram R, Lakshmi M, Venkateswar R, Ravindra P (2007) Economics and environmental impact of bioethanol production technologies: an appraisal. Biotechnol Mole Boil Rev 2(1):014–032

Clean Fuels Development Coalition (2010) Ethanol Fact book. http://cleanfuelsdc.org/wp-content/uploads/2018/05/CFDC-2010-Ethanol-Fact-Book.pdf

Cole B (2006) A Timeline of the Ethanol Industry, Minnesota Public Radio, September 2006. http://minnesota.publicradio.org/projects/2006/09/energyproject/timeline.shtml
Demirbas A (2007) Progress and recent trends in biofuels. Prog Energy Combus Sci 33:1–18
EPA (2011). Biofuels and the environment: first triennial report to congress. Report no. EPA/600/R-10/183F. National Center for Environmental Assessment, EPA
Foody B (1988) Ethanol from biomass: the factors affecting it's commercial feasibility. Iogen Corporation, Ottawa, Ontario, Canada
Gonsalves JB (2006). An assessment of the biofuels industry in India. United nations conference on trade and development, UNCTAD/DITC/TED/2006/6
Green KP (2006). Bringing down gas and oil prices, environmental policy outlook no. 3 (2006). www.aei.org/publication24336/
Hahn-Hägerdal B, Galbe M, Gorwa-Grauslund G, Lidén G, Zacchi G (2006) Bio-ethanol: the fuel of tomorrow from the residues of today. Trends Biotechnol 24:549–556
Hoekman SK, Broch A, Liu X (2018) Environmental implications of higher ethanol production and use in the U.S.: A literature review. Part I-Impacts on water, soil, and air quality. Renew Sustain Energy Rev 81:3140–3158
Jacobson MZ (2007). Effects of Ethanol (E85) versus Gasoline Vehicles on Cancer and Mortality in the United States, Environmental Science & Technology 41, no. 11 (June 1, 2007). http://pubs.acs.org/cgi-bin/sample.cgi/esthag/2007/41/i11/html/es062085v.html
Kim S, Dale EB (2004) Global potential bioethanol production from wasted crops and crop residues. Biomass Bioenergy 26:361–375
Kreider JF, Curtiss PS (2007). Comprehensive evaluation of impacts from potential, future automotive fuel replacements. January 2007 https://doi.org/10.1115/es2007-36234
Langley J (1979). An Offbeat Approach to Alcohol Production, Mother Earth News, July/August 1979. www.motherearthnews.com/Renewable-Energy/1979-07-01/An-Offbeat-Approach-To-Alcohol-Production.aspx
Lechon Y, Cabal H, Saez, R (2005) Life cycle anlysis of wheat and barley crops for bioethanol production in Spain. Int J Agric Resour Gov Ecol 2005 4 (2):113–122
National Academy of Sciences (2007) Water Implications of Biofuels Production in the United States, National Academies Press, Washington, DC, pp 27–36
Padella M, Edwards R, Candela Ripoll I, Lonza L (2018) Estimates of biofuel production costs and cost of savings; JEC Well-to-Tank Version 5, Annex 5; European Commission, Ispra, Italy, JRC115003
Subramanian KA, Singal SK, Saxena M, Singhal S (2005) Utilization of liquid biofuels in automotive diesel engines: an Indian perspective. Biomass Bioenergy 29:65–72
U.S. Department of Energy, Energy Information Administration, "MTBE, Oxygenates, and Motor Gasoline,". www.eia.doe.gov/emeu/steo/pub/special/mtbe
U.S. Environmental Protection Agency, Office of Transportation and Air Quality, EPA Finalizes Regulations for a Renewable Fuel Standard (RFS) Program for 2007 and Beyond, April 2007. www.epa.gov/otaq/renewablefuels/420f07019
Will GF (2008) Survival of the sudsiest. Washington Post, July 10, 2008
Wyman CE, Hinman ND (1990). Ethanol. Fundamentals of production from renewable feedstocks and use as transportation fuel. Appl Biochem Biotechnol 24/25:735–75
https://e360.yale.edu/features/the_case_against_ethanol_bad_for_environment

Further Reading

Argonne National Laboratory. Well-to-wheels greenhouse gas emissions analysis of high-octane fuels with various market shares and ethanol blending levels. Report no. ANL/ESD-15/10; 2015

Cook R, Phillips S, Houyoux M, Dolwick P, Mason R, Yanca C et al (2011) Air quality impacts of increased use of ethanol under the United States' energy independence and security act. Atmos Environ 45(40):7714–7724

Demirbas A (2009) Political, economic and environmental impacts of biofuels: a review. Appl Energy 86:S108–S117

Du C, Zhao X, Liu D, Lin CSK, Wilson K, Luque R, Clark J (2016) Introduction: an overview of biofuels and production technologies. In: Lin CSK, Wilson K, Clark J (eds) Handbook of biofuels production processes and technologies, 2nd edn. Elsevier, Amsterdam, The Netherlands, p 3. ISBN 978-0-08-100455-5

Environment Canada (2007) Comparison of emissions from conventional and flexible fuel vehicles operating on gasoline and E85 fuels. Report no. ERM Report No. 05-039

Environmental assessment of the use of ethanol as a fuel oxygenate: subsurface fate and transport of gasoline containing ethanol (2001) Report to the California State Water Resources Control Board. Report no. UCRL-AR- 145380:2001

Environmental Protection Agency (2011) EPA decision granting partial waiver of clean air act to increase allowable ethanol content to 15%. Federal Register 76(17):4662–3

European Parliament and Council of the European Union (2018) Directive 2018/2001 of the European Parliament and of the Council of 11 December 2018 on the Promotion of the Use of Energy from Renewable Sources (recast). Brussels, Belgium, European Parliament and Council of the European Union

Ghosh P, Ghose TK (2003) Bioethanol in India: recent past and immediate future. Adv Biochem Eng Biotechnol 85:1

Graham LA, Belisle SL, Baas CL (2008) Emissions from light duty gasoline vehicles operating on low blend ethanol gasoline and E85. Atmos Environ 42(19):4498–4516

Green KP (2008). Ethanol and the Environment | American Enterprise Institute. www.aei.org›research-products›report›ethanol-and

Hess P, Johnston M, Brown-Steiner B, Holloway T, de Andrade JB, Artaxo P (2009) Chapter 10: air quality issues associated with biofuel production and use. In: Howarth RW, Bringezu S (eds) Biofuels: environmental consequences and interactions with changing land use. Gummersbach, Germany, pp 169–194

Hill J, Polasky S, Nelson E, Tilman D, Huo H, Ludwig L et al (2009) Climate change and health costs of air emissions from biofuels and gasoline. PNAS 106(6):2077–2082

Hoekman SK, Broch A (2017) Environmental implications of higher ethanol production and use in the U.S.: a literature review. Part II – impacts on biodiversity, ecosystems, land use change, GHG emissions, and sustainability. Renew Sustain Energy Rev. http://dx.doi.org/10.1016/j.rser.2017.05.052

Khattak MA, Mukhtar A, Zareen N, Kazi S, Muhammad Jan M (2016) Green initiatives: a review of ethanol as an alternate automobile fuel. J Adv Rev Sci Res 21(2016):27–42

Kusiima JM, Powers SE (2010) Monetary value of the environmental and health externalities associated with production of ethanol from biomass feedstocks. Energy Policy 38(6):2785–2796

National Renewable Energy Laboratory (2009) Effects of intermediate ethanol blends on legacy vehicles and small non-road engines. Report no. NREL/TP-540-43543, ORNL/TM-2008/117

Renewable Fuels Association (2014) Fueling a nation: feeding the world. The role of the U.S. ethanol industry in food and feed production. Renewable Fuels Association

Runge CF (2016). The case against more ethanol: it's simply bad for environment

Searle S, Pavlenko N, El Takriti S, Bitnere K (2017) Potential Greenhouse Gas Savings from a 2030 Greenhouse Gas Reduction Target with Indirect Emissions Accounting for the European Union; Working Paper 2017-05; International Council on Clean Transportation (ICCT), Washington, DC, USA

Sheehan J, Aden A, Paustian K, Killian K, Brenner J, Walsh M et al (2003) Energy and environmental aspects of using corn stover for fuel ethanol. J Ind Ecol 7(3–4):117–146

Valin H, Peters D, van den Berg M, Frank S, Havlik P, Forsell N, Hamelinck C (2015) The Land use Change Impact of Biofuels Consumed in the EU. Quantification of Area and Greenhouse Gas Impacts, 2015; A Cooperation of Ecofys, IIASA and E4tech. European Commission, Project number: BIENL13120; ECOFYS Netherlands, Utrecht, Netherlands

Williams PRD, Inman D, Aden A (2009) Heath GA. Environmental and sustainability factors associated with Next-generation biofuels in the U.S.: what do we really know?. Environ Sci Technol 43(13):4763–75

Chapter 12
Future of Bioethanol

Abstract Ethanol has experienced unseen levels of attention due to its value as fuel alternative to gasoline, the increase of oil prices, and the climatic changes, besides being a renewable and sustainable energy source, efficient and safe to the environment. Currently, worldwide ethanol production is in high levels. The future of bioethanol appears to be bright as the need for renewable energy sources to replace dependence on foreign oil is in high demand.

Keywords Biofuels · Bioethanol · Fuel ethanol · Sustainable energy source · Renewable energy source

Bioenergy is now becoming a larger part of the global energy mix. There is a prediction that worldwide bioenergy will add up to 20–30% of the overall primary energy by 2035 (IEA 2013). Production of biofuel for transport has shown the most speedy growth, promoted by support from government. For meeting the pushy objectives under the New Policies Scenario, the supply of different kind of biomass needs to increase numerous folds, posing main challenges for agriculture as well as forestry sectors and increasing issues over the potential social-economic and environmental impacts. The technology for production of first generation biothanol is in an advanced state with fully grown technologies, available infrastructure and markets. However, it is condemned for its land use implications on prices of food and production. In the New Policies Scenario, the share of conventional biomass (sugar and starch crops and oil seeds) in total primary energy demand is expected to reduce in the near future (IEA 2013). Conversely, the advanced biofuels obtained from lignocelluloses and algal biomass offers the prospect of increasing biofuels supply with requirement of lesser land whilst increasing alleviation of green house gases. At this stage, the second generation biofuel technologies are fairly advanced, with a small number of commercial scale plants and about 100 plants at pilot scale and demonstration scale globally while the third generation technologies are in the research and development stage. According to IEA (2013), the advanced biofuels would gain market share after 2020 and reach 20% of biofuel supply in 2035. Still, there are some

P. Bajpai, *Developments in Bioethanol*, Green Energy and Technology, https://doi.org/10.1007/978-981-15-8779-5_12

technical and policy barriers to conquer before these technologies can be implemented globally. Higher investment expenses along with higher cost of production make biofuels derived from lignocelluloses less competitive to fossil fuel or several first generation products. Integration of second generation process with first generation infrastructures can be a practical alternative for reducing the investment costs and technological hazards. For achieving reduced production cost, a reliable and unceasing supply of inexpensive feedstock is vital. Moreover, the entire components of the biomass including intermediates and the by-products should also be taken into consideration and used in a biorefinery system for enhancing the economics of the process (opus.lib.uts.edu.au).

Ethanol is receiving attention due to the escalation of oil prices, and the climate change; it is a inexhaustible energy source, efficient and safe to the environment and is a fuel alternative to gasoline. "Currently, worldwide ethanol production is in high levels, and corn is the main raw material used for this purpose, but this scenario may change due to technological improvements that are being developed for production of low cost cellulosic ethanol, as well as for ethanol production from microalgae. It is important to emphasize that, to be a viable alternative, bio-ethanol must present a high net energy gain, have ecological benefits, be economically competitive and able to be produced in large scales without affecting the food provision. The use of various wastes like wood and agricultural wastes) and unconventional raw materials (such as microalgae) can solve the problem without sacrificing food demands" (repositorium.sdum.uminho.pt).

The future of bioethanol appears to be bright as the need for renewable energy sources to replace dependence on foreign oil is in high demand (Basso et al. 2019; Karimi and Chisti 2015; Bajpai 2019; Kumar et al. 2020; Gandham 2018; Maity 2015; Gray et al. 2006; Demirbas 2007; Azadi et al. 2017; Badger 2002; Baeyens et al. 2015; Bajpai 2007; Balat 2011; https://industrytoday.com/the-future-of-ethanol/; Mussatto et al. 2010).

With many nations seeking to reduce petroleum imports, boost rural economies, and improve air quality, world ethanol production has increased significantly in the recent years https://www.globenewswire.com/news-release/2020/04/02/2011181/0/en/Global-Ethanol-Industry.html. The success of domestic ethanol industries in the United States and Brazil has sparked tremendous interest in countries across the globe where nations have created ethanol programs seeking to reduce their dependence on imported energy, provide economic boosts to their rural economies and improve the environment. As concerns over greenhouse gas emissions grow and supplies of world oil are depleted, Europe and countries like China, India, Australia and some Southeast Asian nations are rapidly expanding their biofuels production and use.

A lot of research is being done including turning biomass, materials from plants, into ethanol using special biotechnological methods (Ho et al. 2014). Biomass ethanol is the future of ethanol production because biomass feedstocks, like wheat straw or switchgrass, require less fossil fuels to grow, harvest and produce. It also allows us to utilize more marginal land, such as grasslands, rather than precious acreage devoted to food crops like corn or soybeans. In this way, ethanol production from biomass does not negatively affect the livestock and food industry. The biorefinery, analogous to today's oil refineries, could economically convert lignocellulose to array of fuels and chemicals—not just ethanol by integrating bio—and thermochemical conversion. Fundamental research and partnerships with the emerging bioenergy industry are critical for the success.

There has been continuing research on improving the energy output of ethanol and improvements should keep growing. Right now, more and more E85 stations are popping up everywhere and more products from generators to power tools to lawn mowers will all start to use some alternative fuels. There are already engines that can run 100% pure ethanol and improvements will help migrate these engines to other areas. Big auto manufacturers like Nissan, Ford, and Honda have all invested money into E85 models as well. Portable generators, standby and emergency generators should all start using ethanol as a fuel source as well. Hopefully an alternative fuel like ethanol will be more popular before the big boom in India and China car usage begins to lower environmental pollution (Sindelar and Aradhey 2017).

The emergence of carbon trading programs in response to many countries' ratification of the Kyoto Protocol will also enhance the affordability of ethanol fuels in comparison to gasoline and diesel. Because ethanol fuels offer a substantial reduction in carbon dioxide emissions, users can obtain carbon credits that can be sold to heavy polluters, again reducing ethanol costs while increasing that of fossil fuels. The European Union recently developed a carbon-trading programme while Japan has conducted several scenario simulations, with hopes to initiate its own nationwide trading system. As Russia considers ratification of the Kyoto Protocol, which would bring the agreement into effect, it seems likely that similar carbon trading schemes will continue to emerge around the world.

A combination of well-reasoned government policies and technological advancements in ethanol fuels could guide a smooth transition away from fossil fuels in the transportation sector. As environmental externalities continue to be incorporated into policy consideration and the fledgling industry emerges, ethanol fuels are likely to become an increasing attractive fuel alternative in the foreseeable future. Looking into the future, the ethanol industry envisions a time when ethanol may be used as a fuel to produce hydrogen for fuel cell vehicle applications.

There is a pressing requirement to review existing biofuel policies in a global scenario for protecting the poor and promoting countryside and agricultural development while assuring environmental sustainability. Unquestionably, the interdisciplinary research and development which combine novel molecular strategies for developing strain and process integration in the framework of process engineering will allow expansion and commercialization of pioneering technologies for exploiting the enormous resources. General public is lacking knowledge on biofuels. "Public acceptance for biofuels will be the last challenge to be addressed once all the systems are in place. As public is the main user of fossil fuel for transportation, lack of public support for new transportation fuels can finally lead to disastrous failure" (Masjuki et al. 2013).

Bioethanol will certainly not be enough to match the energy demand that is going to increase because of rapid increase in population and vehicles, and the diminishing natural resources in the world. Bioethanol technology is advancing with each generation, and the production is increasing slowly. The use of renewable raw materials from forest and agricultural wastes, algal biomass, engineered crops and organisms, and tailor-made enzymes acting on multisubstrate could lead to the development of new technologies and thus reduce the current price of bioethanol to a competing price in the global market. This will facilitate longer preservation of natural resources and also considerably reduce environmental pollution. With the development of novel technologies such as genetic manipulation techniques, the future appears bright. Nonetheless, there is a concern for availability of land for growing raw materials and practice crop rotation. The effects on biomass composition also need to be understood correctly. The perception of fourth generation biofuels is exciting and may lead to noteworthy changes in the coming future (www.tandfonline.com; Niphadkar et al. 2017).

"Second generation biorefineries are operating in all regions of the world, bringing far more favorable energy balances to biofuels production than have been previously realized. Substantial displacement of a significant portion of fossil-based liquid fuels has been demonstrated to be a realistic possibility. However, in the face of low petroleum prices, continuing policy support and investment in research and development will be needed to allow biofuels to reach their full potential" (IEA Bioenergy 2013b; Limayem and Ricke 2012; Morales et al. 2015).

There is a lot of requirement for ethanol in the global market. So, it is becoming important to ensure a systematic, continuous supply of inexpensive and abundantly available raw material, for producing fuel ethanol from renewable feedstocks at realistic cost. Developing engineered crops having higher amount of carbon dioxide trapped inside can also help in reducing the cost of ethanol. The cost must be brought down to US$25–50 for making it competitive with fuels derived from petroleum (https://www.iea.org/topics/world-energy-outlook).

For meeting strong demand growth under the New Policies Scenario, the bioenergy supply chain cannot depend only on one source but a blend of different biomass materials including both food and non-food crops. Extensive advancement of the second and third generation biofuel technologies will need reduced costs obtained through further technical progress and a continous policy support. The changeover towards next generation biofuels will offer medium to long lasting solutions to the exhaustion of fossil fuels and international climatic change.

For a booming ethanol industry, support from government is crucial in correcting tax irregularity, exemption from sale tax as well as excise tax, deregulation of raw materials and its price and promising pilot projects and research and development work on bioethanol. Advances in pretreatment by hydrolysis of hemicellulose with an acid catalyst or using consolidated bioprocessing approach with use of novel, customized blend of enzymes for the cellulose breakdown tied with the new development of genetically manipulated microorganism fermenting all possible sugars in biomass to ethanol at a higher productivity are the main important factors for making the bioethanol program successful on a large scale. Another significant aspect by using the bioethanol is its advantage to the environment. Ethanol is one of the best option for fighting vehicular pollution. It burns cleaner. The release of harmful gasses and particulate emissions which cause health problems is reduced. According to Chandel (2007) "The implementation of bioethanol policy can be helpful in improving in environment and rural economic development with sustainable agricultural practices and enhancement of biomass feedstock conscious usage towards the bioethanol industry will bring up the new age farmer into the limelight and horizon of activities and threshold of business to become renewed with options to deal better in life! A better farmer will ultimately usher in a better livelihood for one and all!".

Ethanol produced from lignocelluloses presents a challenge in developing a economically viable technology and with an increasing interest, it is anticipated that such a technology will be developed soon (Dwivedi et al. 2016). The developed technology will be based on the thermochemical platform rather than sugar platform as embedded technologies like gasification and catalytic conversion are already fully matured and only small improvements are required for customizing the current technology for producing ethanol. The advances in reducing the cost of the substrate are expected to help in reducing the total cost of ethanol production. In terms of implementation, only three technologies have achieved wider acceptance. It is necessary to examine other technologies on a commercial scale to understand their potential in meeting the objectives of economically producing ethanol from lignocelluloses (pinchot.org).

"Considering the importance of forest biomass in meeting the cellulosic biomass demand in the near future, more ethanol mills based on different types of forest biomass should be established at numerous levels. This will help in understanding the exact nature of associated problems with the utilization of forest biomass as a feedstock for ethanol production. Cellulosic ethanol holds the promise to supplement the growing energy needs of the nation. However, at the same time, it is important to strike a harmonious chord with the other natural processes that are associated with the production of cellulosic feedstocks. For example, in case of forestry feedstocks, it is important to evaluate the impacts of biomass production on the local biodiversity or on the local watershed. Similarly, in case of agricultural feedstocks, it is important to evaluate the impact on soil and water conservation and nutrient management of soils. Understanding these relationships will help in developing a comprehensive cellulosic feedstock based bioenergy development in the country" (pinchot.org).

References

Azadi P, Malina R, Barrett SRH, Kraft M (2017) The evolution of the biofuel science. Renew Sustain Energy Rev 76:1479–1484

Badger PC (2002) Ethanol from cellulose: a general review. In: Janick J, Whipkey A (eds) Trends in new crops and new uses. ASHS Press, Alexandria VA USA

Baeyens J, Kang Q, Appels L, Dewil R, Lv Y, Tan T (2015) Challenges and opportunities in improving the production of bio-ethanol. Prog Energy Combust Sci 47:60–88 2015

Bajpai P (2007) Bioethanol. PIRA Technology Report, Smithers PIRA, UK Berg C (2004)

Bajpai P (2019) Biomass to energy conversion technologies- the road to commercialization, 1st edn. Elsevier

Balat M (2011) Production of bioethanol from lignocellulosic materials via the biochemical pathway: a review. Energy Convers Manag 52:858–875

Basso TO, Walker RSK, Basso LC, Walker GM (2019) The future of bioethanol. In: Treichel H, Alves Júnior SL, Fongaro G, Müller C (eds) Ethanol as a green alternative fuel: insight and perspectives, pp 259–283. Renewable energy: research, development and policies. Nova Science Publishers, Inc.. https://novapublishers.com/shop/ethanol-as-a-green-alternative-fuel-insight-and-perspectives/

Chandel AK, Chan ES, Rudravaram R, Lakshmi M, Venkateswar R, Ravindra P (2007) Economics and environmental impact of bioethanol production technologies: an appraisal. Biotechnol Mole Boil Rev 2(1):014–032

Demirbas A (2007) Progress and recent trends in biofuels. Progr Energ Combust Sci 33:1–18

Dwivedi P, Alavalapati JRR, Lal P (2016) *Cellulosic ethanol production* in the *United States: Conversion* technologies, current production status, economics, and emerging developments. Energy Sustain Dev 13(3):174–182

Gandham S (2018) "A technical perspective on 2G biofuels and the way forward" for "EU-India Conference on Advanced Biofuels"

Gray KA, Zhao L, Emptage M (2006) Bioethanol. Curr Opin Chem Biol 10:141–146

Ho DP, Ngo HH, Guo W (2014) A mini review on renewable sources for biofuel. Bioresour Technol 169:742–749

IEA Bioenergy (2013) Biofuel-driven biorefineries: a selection of the most promising biorefinery concepts to produce large volumes of road transportation biofuels by 2025. IEA Bioenergy, Netherlands

IEA, International Energy Agency (2013) World Energy Outlook 2013. International Energy Agency, IEA/OECD, Paris

Karimi K, Chisti Y (2015) Future of bioethanol. Biofuel Res J 2(2015):147

Kumar B, Bhardwaj N, Agrawal K, Verma P (2020) Bioethanol Production: generation-based comparative status measurements. In: Srivastava N, Srivastava M, Mishra P, Gupta V (eds) Biofuel production technologies: critical analysis for sustainability. Clean Energy Production Technologies, Springer, Singapore

Limayem A, Ricke SC (2012) Lignocellulosic biomass for bioethanol production: current perspectives, potential issues and future prospects. Prog Energy Combust Sci 38:449–467

Maity SK (2015) Opportunities, recent trends and challenges of integrated biorefinery: part II. Renew Sustain Energy Rev 2015(43):1446–1466

Masjuki HH, Kalam MA, Mofijur M and Shahabuddin M (2013) Biofuel: policy, standardization 523 and recommendation for sustainable future energy supply. Energy Procedia 42:577–586. 524 (2013)

Morales M, Quintero J, Conejeros R, Aroca G (2015) Life cycle assessment of lignocellulosic bioethanol: environmental impacts and energy balance. Renew Sustain Energy Rev 42:1349–1361

Mussatto SI, Dragone G, Guimarães PM, Silva JP, Carneiro LM, Roberto IC, Vicente A, Domingues L, Teixeira JA (2010) Technological trends, global market, and challenges of bio-ethanol production. Biotechnol Adv 28(6):817–30. https://doi.org/10.1016/j.biotechadv.2010.07.001

Niphadkar S, Bagade P, Ahmed S (2017) Bioethanol production: insight into past, present and future perspectives. Biofuels 2017(8):1–10

Sindelar S, Aradhey A (2017). India: Biofuels Annual, 2017; USDA Gain Report # IN7075

pinchot.org

https://industrytoday.com/the-future-of-ethanol/

https://www.globenewswire.com/news-release/2020/04/02/2011181/0/en/Global-Ethanol-Industry.html

https://www.tandfonline.com/

https://www.iea.org/topics/world-energy-outlook

Index

P. Bajpai, *Developments in Bioethanol*, Green Energy and Technology,
https://doi.org/10.1007/978-981-15-8779-5

The manufacturer's authorised representative in the EU is Springer Nature Customer Service Centre GmbH, Europaplatz 3, 69115 Heidelberg, Germany. If you have any concerns regarding our products, please contact ProductSafety@springernature.com

Printed and bound by CPI Group (UK) Ltd, Croydon, CR0 4YY
30/09/2026
02225246-0008